ISBN 978-1-332-27559-5
PIBN 10307724

1 MONTH OF
FREE
READING

at
www.ForgottenBooks.com

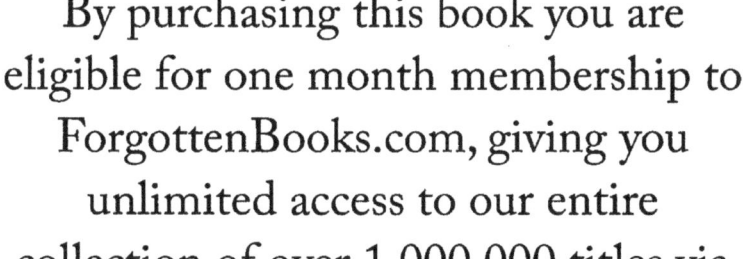

By purchasing this book you are eligible for one month membership to ForgottenBooks.com, giving you unlimited access to our entire collection of over 1,000,000 titles via our web site and mobile apps.

To claim your free month visit:

www.forgottenbooks.com/free307724

English
Français
Deutsche
Italiano
Español
Português

www.forgottenbooks.com

Mythology Photography **Fiction**
Fishing Christianity **Art** Cooking
Essays Buddhism Freemasonry
Medicine **Biology** Music **Ancient**
Egypt Evolution Carpentry Physics
Dance Geology **Mathematics** Fitness
Shakespeare **Folklore** Yoga Marketing
Confidence Immortality Biographies
Poetry **Psychology** Witchcraft
Electronics Chemistry History **Law**
Accounting **Philosophy** Anthropology
Alchemy Drama Quantum Mechanics
Atheism Sexual Health **Ancient History**
Entrepreneurship Languages Sport
Paleontology Needlework Islam
Metaphysics Investment Archaeology
Parenting Statistics Criminology
Motivational

A BOOK ON BUILDING, Civil and Ecclesiastical; including Church Restoration; with the Theory of Domes and the Great Pyramid, and Dimensions of many Churches and other Great Buildings. By Sir EDMUND BECKETT, Bart., LL.D., Q.C., F.R.A.S., Chancellor and Vicar-General of York. Second Edition, Enlarged, 4s. 6d.; cloth boards, 5s.

PLUMBING: a Text-book to the Practice of the Art or Craft of the Plumber. With Chapters upon House Drainage, embodying the latest improvements. By W. P. BUCHAN, Sanitary Engineer. Second Edition, revised and much enlarged. With 300 Illustrations. 3s. 6d.; cloth boards, 4s.

HOUSE PAINTING, GRAINING, MARBLING, AND SIGN WRITING: a Practical Manual of. With 9 Coloured Plates of Woods and Marbles, and nearly 150 Wood Engravings. By ELLIS A. DAVIDSON. Second Edition, carefully revised. 5s.; cloth boards, 6s.

THE ART OF LETTER PAINTING MADE EASY. By JAMES GREIG BADENOCH. Illustrated with 12 full-page Engravings of Examples. 1s.

HINTS TO YOUNG ARCHITECTS. By GEORGE WIGHTWICK. New and Enlarged Edition. By G. HUSKISSON GUILLAUME, Architect. With numerous Woodcuts. 3s. 6d.; cloth boards, 4s.

THE DRAINAGE OF TOWNS AND BUILDINGS, By G. DRYSDALE DEMPSEY, C.E. 2s. 6d.

THE DRAINAGE OF DISTRICTS AND LANDS. By G. DRYSDALE DEMPSEY, C.E. Illustrated. 1s. 6d.

** With " Drainage of Towns and Buildings," in one vol. 3s. 6d.

THE BLASTING AND QUARRYING OF STONE, for Building and other purposes, and on the Blowing up of Bridges. By Gen. Sir JOHN BURGOYNE, Bart., K.C.B. Illustrated. 1s. 6d.

COTTAGE BUILDING; or, Hints for Improved Dwellings for the Labouring Classes. By C. BRUCE ALLEN, Architect. Illustrated. New Edition, 1s. 6d.

FOUNDATIONS AND CONCRETE WORKS. By E. DOBSON, M.R.I.B.A., &c. 1s. 6d.

LIMES, CEMENTS, MORTARS, CONCRETES, MASTICS, PLASTERING, &c. By G. R. BURNELL, C.E. 1s. 6d.

WARMING AND VENTILATION; being a Concise Exposition of the General Principles of the Art of Warming and Ventilating Domestic and Public Buildings, Mines, Lighthouses, Ships, &c. By CHARLES TOMLINSON, F.R.S., &c. Illustrated. 3s.

CONSTRUCTION OF DOOR LOCKS AND IRON SAFES. Edited by CHARLES TOMLINSON, F.R.S. 2s. 6d.

CROSBY LOCKWOOD & CO., 7, STATIONERS' HALL COURT, E.C.

JOHN A. SEAVERNS

A PRACTICAL·TREATISE ON

COACH-BUILDING

HISTORICAL AND DESCRIPTIVE

CONTAINING

FULL INFORMATION OF THE VARIOUS TRADES AND
PROCESSES INVOLVED, WITH HINTS ON THE
PROPER KEEPING OF CARRIAGES, &c.

WITH FIFTY-SEVEN ILLUSTRATIONS

By JAMES W. BURGESS

LONDON
CROSBY LOCKWOOD AND CO.
7, STATIONERS' HALL COURT, LUDGATE HILL
1881

LONDON:
PRINTED BY J. S. VIRTUE AND CO., LIMITID,
CITY ROAD.

PREFACE.

IT is singular that such an important industry as coach-building should have received such slight attention from writers either at home or abroad; yet such is the case, the last book dealing in any way exhaustively with the trade having been published some fifty years ago. The manufacturers themselves have doubtless very copious notes, which, if printed, would make several large volumes; but they do not publish the result of their experience to the world, and consequently the general public, and more particularly apprentices and others whose occupations or amusements may be in any way connected with the trade, have no means of gaining an insight into this branch of industrial art.

It is hoped that this book will supply, to some extent at least, this deficiency. Its object is general utility rather than technical instructions on minor details. The principles on which carriages ought to be constructed, rather than the arbitrary proportions of parts, are what the author has sought to make clear.

From an antiquarian point of view the history of carriages is very interesting. In the first chapter the gradual development of vehicles and their parts from the first rude forms of raft and sledge, down to the

shapes with which we are now familiar, has been carefully dealt with, and all matters of interest in connection therewith have been added.

In dealing with the practical part of the business, the
preliminary operations of the carriage designer and
draughtsman are first considered, and directions given
how to make the drawing of a proposed vehicle; from
whence the reader is taken into the workshop, where he
is introduced to the points which usually have to be
considered before commencing the construction of a
vehicle, and instructions given as to the preparation of
the full-sized working draughts or drawings. The
component parts of body and carriage are then par-
ticularised, and notes added to assist in their being
properly understood. A glance is given at the various
materials used, and then the most important parts of
the vehicle are dealt with separately, as wheels, axles,
springs, &c., both the theory and manufacture of each
part being given, in order to make clear the work it
has to perform in operation, and the means of obviating
difficulties.

In the chapter devoted to the painting department a
slight sketch has been given of the theory of colour,
in order that the unscientific reader may understand
the harmonies and contrasts of colour, which are so
important to the successful conduct of this branch.
The materials and implements that the painter uses are
described, and the vehicle followed through the successive stages of priming, painting, colouring, and
varnishing. In addition to this a chapter is given on

" Ornamental Painting," *i.e.* painting monograms, crests, and coats of arms on the carriage panels.

In the chapter " General Observations on the Trade " several interesting facts and statistics have been dealt with ; such as the present position of carriage artisans ; application of machinery ; the coach trade in America and in India ; also the theory of lengthening or shortening the carriage parts.

A chapter on " Invention " has been added, which it is trusted may prove of value. As a concluding chapter, some hints on the proper treatment of carriages by their private owners have been given, which will no doubt be useful to the coach-builder.

What few authorities there are on the subject have been carefully consulted by the author, with a view to making the book as accurate and reliable as possible.

If any member of the trade desires to make suggestions as to improvement in any of the processes given, his communication, addressed to the author (care of the publishers), will be carefully considered in connection with subsequent editions of the work.

J. W. B.

London, *April*, 1881.

CONTENTS.

A PRACTICAL TREATISE

ON

COACH-BUILDING.

CHAPTER I.

GENERAL HISTORY.

THE origin of the word coach has not yet been accurately determined. Menage says it is taken from the Latin *vehiculum*, which most people will take the liberty of doubting; Wachten, from the German *kutten*, to cover; Lye, from the Belgic *koetsen*, to lie along, or, as it really means, a couch or chair; it has also been tried to prove that the word is of Hungarian origin, and that it took its name from *Kotsee*, the old name of the province of Wiesellung, where various kinds of carriages were made; and in Beckmann's "History of Inventions" it is mentioned that "when the Archbishop received certain intelligence that the Turks had entered Hungary, not contented with informing the king by letter, he speedily got into one of those light carriages of the place they call *kotcze*, and hastened to his majesty." This, in addition to the fact that some years previously the King of Hungary presented to the Queen of Bohemia a vehicle that excited great wonder and admiration, by reason of its trembling (*branlant*), showing clearly that it must have been suspended, is strongly in favour of the

B

Hungarian coachmakers; but we must leave it to the philologists to determine the exact truth, for what with the *caroche* of France, the *caroce* of Italy, the *carri-cochè* of Spain, and our own coach, the head gets somewhat bewildered, and is fain to take refuge in the simple *carruca* of ancient Rome, from which these appellations most probably had their rise. In any case the honour must be a divided one, as the *caretta*, *chare*, car, *charat*, &c., must have been the earliest forms of the derivation, as such were the names given to the first vehicles; later, we have the Hungarian *kotcze*, the German *kutsche*, &c., and adding both form and name to what had gone before, produced a mixed vehicle with a mixed appellation. Dr. Johnson defines coach as a "little carriage." The *large* carriage that he had in his mind's eye at the time must have been a marvellous vehicle.

The progress of the art of coach-making, like the progress of most inventions and discoveries, has been rather slow, we may say remarkably slow; sometimes it made a sudden start, but a reaction in the other direction generally settled it before much advance had been made; but seeing that the early portions of the Old Testament contain references to wheel carriages, it does seem rather strange that perfection should take so long to arrive at. This may be partly accounted for by the fact that the nations of the earth were always at war with one another, and consequently had no time to foster inventive power. And this has unfortunately been the case until comparatively recent times.

The first land carriages were doubtless very primitive contrivances. Though the "chariot" and the "waggon" are mentioned in Genesis, no description is given of their construction. Joseph rode in the second chariot of Pharaoh as a mark of great honour and dignity. "Waggons" were dispatched from the court of Egypt to convey thither the wives and little ones of the family of Jacob. From this, as well as the fact of the brethren of Joseph bearing their corn

away on asses, we may infer that wheel carriages, even of
the most simple construction, were not in general use at this
time. It is very probable that the common vehicle of the
period was an embryo sledge, drawn by man or beast along
the ground.

The Bible and the hieroglyphics on the various ruins of
ancient Egypt furnish us with the earliest authentic records.
In the case of Egypt this is particularly valuable to us,
because of the great degree of culture arrived at in the
civilised arts. In fact it is the chief country of which we
have any record of the progress of these arts, and though
not actually established, it is extremely probable that to the
Egyptians we owe the invention, or at least the introduction
of the wheel. These people were early engaged in the
erection of large buildings and monuments, of which the
pyramids and sphinxes are such striking examples ; and in
order to convey the enormous blocks of stone and granite to
their ultimate destination, the roller would be the first thing
to suggest itself as a means of facilitating transit. The next
step would be the formation of a truck, to which these rollers
could be attached, and on which could be placed the materials
to be moved. Progression with a contrivance of this kind
would necessarily be rather slow, but it would soon become
apparent that if a larger roller were used the motion could
be accelerated. The next improvement would be an en-
deavour to lighten the rollers by sawing them into thick
slices, and connecting them by a horizontal roller of smaller
dimensions, giving a rude representation of a wheel and
axle. The agricultural carts used by the peasantry of Chili,
in South America, were made in this fashion until very
recently. The further lightening of these cars would follow
almost as a matter of course, by cutting the slices of the
trunk to form the wheel, thinner, and further by cutting
away portions of this slice, forming spokes. The wheel
having arrived at this stage of perfection, the axle would

call for a little attention.   Up till the present, they would be
fixed firmly to the wheels and revolve with them.   This
arrangement would cause great inconvenience in turning, for
one wheel would revolve more rapidly than the other, by
reason of the circle described by one wheel in turning round
being greater than that of the other, and the vehicle would
be liable to overturn.   The next step was to arrange that
the wheels could revolve independently of the axle.   This
being done, we have the wheel, in its principles, the same
as at present.

The paintings and sculptures upon the walls of the temples

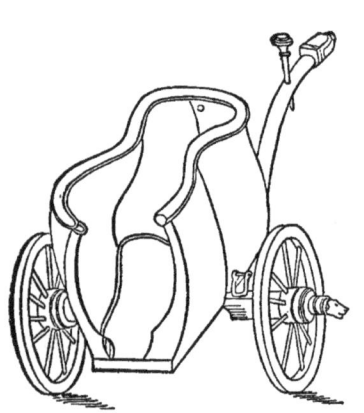

Fig. 1.—Egyptian Chariot.

Fig. 2.—Egyptian Chariot.

and tombs of Egypt show that wheeled carriages were in use in
that country at an early period (Figs. 1 and 2).   In the Bible
they are usually translated "chariot."   They are of great
interest to us, as they formed the chief means of conveying man
for 2,000 years before Christ, and were more or less the type
of all the other vehicles of the ancient world.   We find certain
words used in describing them, both by Homer, who lived
1,000 years B.C., and by Moses, who lived at least 500 years
earlier ; and these words are the technical terms in use at the
present day, such as axles, wheels, naves, tyres, spokes, &c.
It is reasonable to infer from this, that the art to which

these terms apply must have existed prior to the writers' description ; so that any doubt as to the correctness of the Egyptian sculptures must be dispelled by the references of the above authors. In the fifth book of the Iliad " The awful Juno led out the golden-bitted horses, whilst Hebe fitted the whirling wheels on the iron axle of the swift chariot. The wheels had each eight brazen spokes, the felloes were of gold, secured with brazen tyres all round, admirable to the sight. The seat was of gold hung by silver cords, the beam or pole was of silver, at the end of which were hung the golden yoke and the golden reins."

The car was greatly used by the Romans, being adopted from the one used by the Etrurians (Fig. 3), a neighbouring country on the Italian peninsula. These latter people were tradition-ally the first to place a hood or awning over the open two-wheeled car, and they showed great taste in decorat-ing their vehicles in the manner familiar to

Fig. 3.—Roman Chariot.

us by the remains of their pottery. A very fine copy of one of the Roman cars is in the museum at South Kensington, cast from the original in the Vatican.

Herodotus (450 B.C.) mentions that the Scythians used a vehicle which consisted of a rough platform upon wheels, on which was placed a covering like a beehive, composed of basket work and covered with skins. When they pitched anywhere these huts were taken off, and served them as dwellings in lieu of tents. Fig. 4 shows one of their chariots.

The war chariots used by the Persians were much larger than those used by contemporary nations. The idea seems

to have been to form a sort of turret on the car to protect
the warriors in action.   These vehicles were provided with
curved blades, like scythes, which projected from the axletrees, for the purpose of maiming the enemy as they drove
through them.

At the period of the invasion of this country by the
Romans, a car or
chariot seems to have
been in use which
they had not met
with before.   It was
larger than the Ro-
man car, and pos-
sessed a seat, from
which feature it was
called *essedum*.   It

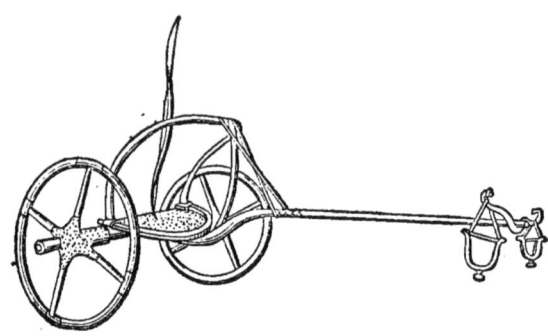

Fig. 4.—Scythian Chariot.

was doubtless an improved vehicle of its kind, for Cicero,
writing to a friend in Britain, says " that there appeared
little worth bringing away from Britain except the chariots,
of which he wished his friend to bring him one as a pattern."

Sir William Gell, in his work on Pompeii, which was
destroyed A.D. 79, mentions that three wheels had been dug
out of the ruins in his day, very much like our modern
wheels—a little dished, and 4 feet 3 inches high, with ten
spokes rather thicker at each end than in the middle.
He also gives an illustration of a cart used for the conveyance of wine in a large skin or leather bag; it is a four-
wheeled cart, with an arch in the centre for the front wheel
to turn under.   The pole appears to end in a fork, and to be
attached to the axle bed.

On the decline of the Roman power, many of the arts of
civilisation which they had been instrumental in forwarding
fell into disuse.   The skilled artisans died and left no suc-
cessors, there being no demand for them.   This will account
for no mention being made of carriages or chariots for some

centuries. Of course there were various primitive contrivances in use to which the name of cart was given, but the great and wealthy moved about the cities or travelled on horseback, or if they were incapable of this, they used litters carried by men or horses. The great bar to the general adoption of wheeled carriages was undoubtedly the very bad state of the roads.

An evident improvement in construction was made by the Saxons. In the Cotton Library there is a valuable illuminated manuscript, supposed to be the work of Elfricus, Abbot of Malmesbury. The subject is a commentary on the Books of Genesis and Exodus, with accompanying illustrations. In one of these is represented the first approach to a slung carriage ; and it may be interesting to the lovers of historical coincidence that it is given in an illustration of the meeting of Joseph and Jacob, and in that part of the Bible which first makes mention of vehicular conveyance. The chariot in which Joseph is seated is a kind of hammock (most probably made of leather, which was much used by the Anglo-Saxons), suspended by iron hooks from a framework of wood. It moves upon four wheels, the construction of which is not clear, owing to the decorative license taken with them by the artist. The father of Joseph is placed in a cart, which we doubt not, from its extreme simplicity, is a faithful type of those of the time. This proves the illuminator to have been true to his subject and the custom of the period in which he lived, as the chariot was monopolised by the great men, while the people rode in carts.

With the Normans came the horse litter, a native originally of Bithynia, and from thence introduced into Rome, where it is still used by the Pope on state occasions, and also among the mountain passes of Sicily, as well as in Spain and Portugal. Malmesbury records that the dead body of Rufus was placed upon a *rheda caballaria,* a kind of horse litter. King John, in his last illness, was conveyed from the Abbey of Swinstead *in lectica equestre.* These were for several succeeding reigns the only carriages in use for persons of distinction. Froissart writes of Isabel, the second wife of Richard II., as " La june Royne d'Angleterre en une litieré moult riche qui etoit ordonée pour elle." These litters were seldom used except on state occasions. When Margaret, daughter of Henry II., went into Scotland, she is described as journeying on a " faire palfrey," but after her was conveyed by two footmen " one very riche litere, borne by two faire coursers ¹vary nobly drest; in which litere the sayd queene was borne in the intryng of the good towns or otherwise to her good playsher."

Carriages proper were first introduced on the continent. Italy, France, Spain, and Germany contend with each other for the honour of the first introduction. The earliest record we have is on the authority of Beckmann, who says that, when at the close of the thirteenth century Charles of Anjou entered Naples, his queen rode in a *caretta,* the outside and inside of which were covered with sky-blue velvet interspersed with golden lilies.

The English were not long before they adopted this new innovation. In an early English poem called the " Squyr of Low Degree," supposed to be before the time of Chaucer, the father of the Princess of Hungary thus makes promise :—

> " To-morrow ye shall on hunting fare,
>   And ride my daughter in a *chare.*
>   It shall be covered with velvet red,
>   And cloths of fine gold all about your head,
>   With damask white and azure blue,
>   Well diapered with lilies new :
>   Your pomelles shall be ended with gold,
>   Your chains enamelled many a fold."

The pomelles were doubtless the handles to the rods affixed towards the roof of the " chariette," and were for the purpose of holding by when deep ruts or obstacles in the roads caused an unusual jerk in the vehicle.

On the continent, there seems to have been a great deal of opposition to the use of carriages. In 1294, Philip, King of France, issued an ordinance prohibiting the citizens' wives the use of cars or chars; and later on, Pope Pius IV. exhorted his cardinals and bishops not to ride in coaches, according to the fashion of the time, but to leave such things to women; and it really was thought *infra dig.* for a man to travel other than on horseback. Even his Holiness the Pope rode upon a grey horse; though to indemnify him for the exertion, his horse was led, and his stirrup held by kings and emperors.

These exhortations had about the same effect as James I.'s " Counterblast to Tobacco ; " they created an increased demand, and the people showed their sense in preferring the ease that does no injury to the self-denial that does no good, in spite of the opposition of their superiors.

The first coach made in England was for the Earl of Rutland, in 1555, and Walter Rippon was the builder. He afterwards made one for Queen Mary. Stow's " Summerie of the English Chronicle " is the authority upon which this statement is made.

In a postscript to the life of Thomas Parr, written by Taylor, the Water Poet (and a mortal enemy to land carriages), we find the following note : " He (Parr) was eighty-one years old before there was any coach in England (Parr was born in Edward IV.'s reign in 1483); for the first ever seen here was brought out of the Netherlands by one William Boonen, a Dutchman, who gave a coach to Queen Elizabeth, for she had been seven years a queen before she had any coach; since when they have increased with a mischief, and ruined all the best housekeeping, to the undoing of the watermen, by the multitudes of Hackney coaches. But they never swarmed so much to pester the streets as they do now till the year 1605 ; and then was the gunpowder treason hatched, and at that time did the coaches breed and multiply." Taylor is to be thanked, not only for his information, but for his capital though unconscious burlesque upon those fancied philosophers who talk of cause and effect, where events, because they happen in sequence, are made to depend one on the other, when the fact of their being two things apart makes them independent existences.

We have not space to dwell upon these old specimens at length. Queen Elizabeth's coach is called by an old author "a moving temple." It had doors all round, so that when the people desired, and the virgin queen was agreeable, they might feast their eyes on the beauty of its trimming or linings.

The following entry in Sir William Dugdale's diary may be interesting: "1681. Payd to Mr. Meares, a coach-maker in St. Martin's Lane, for a little chariot wch I then sent into the country, £23 13s. 0d., and for a cover of canvas £01 00s. 00d.: also for harness for two horses £04 00s. 00d."

The opposition on the part of the watermen to the introduction of coaches assumed rather serious proportions, more especially as the populace sided with them ; to such a height did the antagonism run that a movement was made to introduce a Bill into Parliament to prevent the increase of coaches; the apology for its introduction being, that in war time it would be a matter of great difficulty to mount the troops if so many horses were monopolised for these coaches. Luckily, however, it came to nothing, and the antipathy gradually died out.

Coaches and vehicles of all descriptions now became general, and in 1635 a patent was granted to Sir Saunders Duncombe for the introduction of sedans ; their purpose being " to interfere with the too frequent use of coaches, to the hindrance of the carts and carriages employed in the necessary provision of the city and suburbs." A rivalry now sprung up between coach and sedan, and gave rise to a humorous tract, in which they hold a colloquy as to which should take precedence, a brewer's cart being appointed umpire.

The coaches at this period were fearfully and wonderfully made. There are several examples of them scattered about in the various museums. The people who used them at this time had no great ideas of them, for so formidable an affair was the undertaking of a journey reckoned, that even from Birmingham to London a departure was the signal for making a will, followed by a solemn farewell of wife, children, and household !

Towards the end of the seventeenth century improvements began to take place. In Wood's diary mention is made of a machine called the " Flying Coach," which performed the journey between Oxford and London in thirteen hours! This was express rate for that age, especially as there was some talk of making a law to limit the ground covered by a coach to thirty miles a day in summer, and twenty-five miles a day in winter. Oh, those good old times ! The outcry lessened, and the imperfect vehicles and bad roads were left to passengers unmolested. What the latter were may be imagined from the fact that, when Charles III. of Spain visited England, and Prince George of Denmark went out to meet him, both princes were so impeded by the badness of the roads that their carriages were obliged to be borne on the shoulders of the peasantry, and they were six hours in performing the last nine miles of their journey.

In the eighteenth century improvements were made in the construction of coaches, but they were still heavy lumbering contrivances, so that little or no progress was made in the rate at which they travelled. Even so late as 1760 a journey from Edinburgh to London occupied eighteen days, a part of the roads being only accessible by pack horses. There is

a very good specimen of the vehicle of the early part of the
eighteenth century in the South Kensington Museum, belonging to the Earl of Darnley's family, and is well worthy of study
as being one of the lightest examples known of this period.

In the Museum of South Kensington is also an excellent
example of the fully developed coach of 1790. It is a very
massive-looking affair, and belonged to the Lord Chancellor
of Ireland; it looks very much like a faded edition of the
City state coach now, though when new it doubtless had a
very good appearance. It consists of a very large body,
suspended from upright or whip springs by means of leather
braces; the standing pillars slope outwards, making the sides
longer at the roof than at the elbow line. The wheels are of
good height, and the carriage part is very massively constructed, the upper part being finished off with scroll ironwork, and on this in the front the coachman's hammercloth
is raised. The panels are painted with landscapes, &c., by
Hamilton, R.A., and no doubt altogether it cost a deal of
money.

Vehicles now began to assume that variety of shape and
form of which we have in our own time so many specimens.
There were Landaus, introduced from a town of that name in
Germany ; these were, like the coaches, only made to open in
the centre of the roof just as they do now, but instead of the
covering falling into a horizontal line it only fell back to an
angle of 45 degrees, and this pattern was maintained for a
number of years. Landaulets were chariots made to open.
Generally speaking, the difference between a coach and a
chariot was that the former had two seats for the accommodation of passengers, and the latter but one, and in appearance was like a coach cut in half. Then came phaetons,
barouches, sociables, curricles, gigs, and whiskies, which, in
their general form and attributes, were similar to the vehicles
of the present day which bear these names. In those days
fast driving was all the " go," and young men vied with each

other in driving the loftiest and most dangerous gigs and phaetons. Contemporary literature teemed with romantic tales of spills and hairbreadth 'scapes from these vehicles, and yet dilated on the fearful pleasure there was in driving them.

The larger wheeled vehicles were hung upon framed carriages, with whip springs behind and elbow springs in front, like the gentlemen's cabriolets of the present day. When drawn by two horses they were called curricles, or by one horse, chaises. There was a little variation in the shape of the body, viz. the full curricle pattern and the half curricle, with or without a boot, similar to a Tilbury or a gig body. The wheels were 4 feet 8 inches to 5 feet in height. Lancewood was then used for shafts.

It is at the beginning of the nineteenth century that real progress is to be found in coaches and other carriages. In 1804, Mr. Obadiah Elliott, a coachmaker of Lambeth, patented a plan for hanging vehicles upon elliptic springs, thus doing away with the heavy perch, as the longitudinal timber or iron connecting the hind carriage with the fore carriage is called. Perches are still used, but are chiefly confined to coaches proper, or those hung upon C springs. Elliott also considerably lightened the carriage part of the vehicles he turned out. This was the first step to a grand revolution in the manufacture of carriages, which was to affect every variety of vehicle, great or small. Elliott's enterprise was rewarded by the gold medal of the Society of Arts, and by his business becoming a very prosperous one, for the public were not slow in discovering the advantages arising from great lightness in vehicles.

A print, published in 1816, shows a landaulet hung on elliptic springs, four in number, with a square boot framed to the body, and the driving seat supported on ironwork high above the boot. Behind there is a footboard supported on the pump-handles. The distance between the axletrees is very short, only 6 feet 6 inches from centre to centre.

The body is rather small, and the wheels are 3 feet 8 inches and 4 feet 8 inches high respectively, and the bottom of the body is 3 feet 6 inches above the ground. The span or opening of the springs is 10 inches.

In 1814 there were 23,400 four-wheeled vehicles paying duty to Government, 27,300 two-wheeled, and 18,500 tax-carts in Great Britain, showing a total of 69,200 vehicles. The later returns will show how much a reduction in the duties and the use of elliptic springs have promoted the increase of vehicles of all kinds.

A vehicle much in fashion at this period was the curricle, which had been in use for some time in Italy, where it was suspended from leather braces. Springs were added by the French, and, on its being introduced here, the English altered the shape, giving the back a graceful ogee curve, improved the hood, and added a spring bar across the horses' backs. It was a vehicle of easy draught, and could be driven at great speed. Unfortunately it was rather dangerous if the horse shied or stumbled, and this tended to reduce the demand for it, and it was gradually superseded by the cabriolet, though Charles Dickens used one as soon as he could afford it, and Count D'Orsay had one made as late as 1836.

The vehicle called the briska, or britchka, was introduced about 1818 from Austria. It was hung both upon C springs and elliptic springs, and was made in various sizes for different requirements. It was nearly straight along the bottom. The hind panel was ogee shaped, and the front terminated in a square boot. There was a rumble behind, and the back seat was fitted with a hood which could be raised or lowered at pleasure, and the knees were covered by a folding knee flap. This was an inconvenient vehicle for our climate, as only half the number could be sheltered in wet weather that could be accommodated in dry. It was very fashionable for a time, but died out about 1840.

The " Stanhope " takes its name from being first built to the order and under the superintendence of the Hon. Fitzroy Stanhope, by Tilbury, the builder of the vehicle bearing that name. It was shaped like the old ribbed gig, but was hung upon four springs, two of which were bolted between the shaft and axle, and the other two crossways, parallel to the axle at either end of the body, and shackled to the side springs. Stanhopes are an easy kind of vehicle, and do not rock so much as other gigs behind a rough-trotting horse. At the same time they are rather heavy, owing

Fig. 5.—Stanhope.

to the large amount of iron plating used to strengthen the shafts, &c.

The " Tilbury " was very much like the Stanhope, but had no boot, and like it was heavily plated with iron. It was hung by two elbow springs in front, with leather braces to the shafts or front cross bar, and behind by two elbow springs passing from beneath the seat to a cross spring raised to the level of the back rail of the body by three straight irons from the hind part of the cross bar. Later, two more springs were added between the axletree and the shafts, by scroll irons. The Tilbury was a very good-looking and durable vehicle, but its weight took away the public favour, and it went out of fashion about 1850. It was, however, adopted with great success by Italy and other continental countries, where the roads are bad, and solidity of construction is the first consideration.

Dog-carts and Tandem-carts are too well known to need description. The former were so called from their being used for the conveyance of sporting dogs, such as greyhounds or pointers, and the slats or louvre arrangement of the sides was for the purpose of admitting air to the animals;

Fig. 6.—Tilbury.

though scarcely ever used for this purpose now, the original plan has been pretty closely adhered to, except that the boot is considerably reduced and made to harmonise more with the other parts.

Some of the greatest improvements in the shape and style of various vehicles were effected by a celebrated maker named Samuel Hobson, who remodelled and improved pretty nearly every vehicle which came under his hands. He particularly directed his attention to the true proportion of parts, and artistic form of carriages. He lowered the bodies, and lengthened the under or "carriage" part. The curves and sweeps also received due attention. In fact, he carefully studied those "trifles" (as Michael Angelo's friend would have termed them) on which depended the success of the production as a work of art. Imitation being the sincerest form of flattery, the other coachmakers soon showed their sense by copying his best ideas, though, to give these other coachmakers their due, they greatly assisted Mr. Hobson with suggestions for improvements, and as a reward availed themselves of his superior talent for working on these ideas.

As our interior trade and manufactures increased, the custom arose of sending commercial travellers throughout England to call attention to the various goods, and it was found very convenient to send these travellers in light vehicles which could convey samples of the various articles. This led to a very great increase in the number of gigs; and about 1830 one coach factory of London supplied several hundreds of these vehicles to travellers at annual rentals. And though on the introduction of the railway system long journeys by road were unnecessary, these gigs were found of great use in town and suburban journeys, and in London they may be seen by hundreds daily, and they are scarcely used by any one else but commercial travellers. They are too familiar to need detailed description.

In 1810 a duty was levied by Government upon vehicles for sale. It was repealed in 1825, but the returns give the number of vehicles built for private use in 1814 as 3,636, and in 1824 as 5,143, whilst the number of carriages in use in 1824 had grown to 25,000 four-wheeled, and 36,000 two-wheeled, besides 15,000 tax-carts; an increase since 1814 of 20,000 vehicles.

In 1824 there was built for George IV. a low phaeton, called a pony phaeton, which has since become very common, and has undergone but very little change from the original. It was a cab shape, half-caned, with a skeleton bottom side hung upon four elliptical springs, with crane ironwork back and front. It was drawn by two ponies; the wheels were only 21 and 33 inches high.

A carriage had been introduced from Germany, called a droitska or droskey—an open carriage with a hood, on a perch, and suspended from C springs. The peculiarity was, that the body was hung very near the perch, so that the seat was only 12 inches above the hind axletree, and the place for the legs was on either side of the perch. The chief merits of this vehicle consisted in its lightness as compared with barouches and briskas, and its shortness.

The cab phaeton was invented by Mr. Davies, of Albany Street, about 1835; it consisted of a cab body with a hood, hung upon four elliptic springs, and a low driving seat and dasher, for one horse. It met with great success and was soon in general use. It was introduced on the continent, where it became known under the name of " Milord," and became the common hack carriage, after which it went out of fashion with the upper circles. It has, however, been recently revived under the name of " Victoria." The Prince of Wales and Baron Rothschild set the fashion by using Victorias about 1869, and it really is a very elegant and useful vehicle.

In 1839 the first Brougham was built by Mr. Robinson, of Mount Street, for Lord Brougham, since when this has become the most common and the most fashionable vehicle in use. The size of the first brougham was in its chief dimensions similar to those now manufactured; it was hung on elliptic springs in front, and five springs behind. Coach-makers seemed to have lavished the greatest care and attention on these vehicles, in order to turn out the lightest, and at the same time the most artistic contrivance, and great success has attended their efforts.

The foregoing is a brief history of vehicular conveyances from the earliest times to the present. During the last ten or fifteen years many further improvements have been added, tending to produce more perfect vehicles in every respect; but these improvements have been more in matters of detail than those at the commencement of the century, and hence are more likely to escape ordinary observation; but the critical eye will soon discover these changes, and marvel at the short space of time in which the *real* work has been done.

A glance at public carriages may not be out of place. Hackney coaches were first used in England in 1605. These were similar to the coaches used by fashionable people, but they did not ply for hire in the streets, but remained at the hiring yards until they were wanted. Their number soon increased, owing to there being a greater number of persons who wished to hire than could afford to keep a conveyance of their own. In 1635 the number was limited to fifty, but in spite of the opposition of the King they continued to increase in number, and in 1640 there were 300 in London. In Paris they were introduced by Nicholas Sauvage, who lived in a street at the sign of St. Fiacre, and from this circumstance hackney carriages are called "fiacres" in France. In 1772 the hire of a fiacre in Paris was one shilling for the first hour and tenpence for the second. There were 400 hackney coaches in London in 1662, and the Government then levied a yearly duty of £5 each upon them. In spite of this their number had in 1694 increased to 700, a substantial proof of their usefulness.

In 1703 a stage coach performed the journey from London to Portsmouth, when the roads were good, in fourteen hours. From this time there was a gradual increase in the number and destinations of stage coaches.

In 1755 stage coaches are described as being covered with dull black leather, studded with broad-headed nails by way of ornament, and oval windows in the quarters, with

the frames painted red. On the panels the destination of
the coach was displayed in bold characters. The roof rose
in a high curve with a rail round it. The coachman and
guard sat in front upon a high narrow boot, sometimes
garnished with a hammercloth ornamented with a deep
fringe. Behind was an immense basket supported by iron
bars, in which passengers were carried at a cheaper rate
than in other parts of the vehicle. The wheels were painted
red. The coach was usually drawn by three horses, on the
first of which a postillion rode, dressed in green and gold,
and with a cocked hat. This machine groaned and creaked
as it went along, with every tug the horses gave, though
the ordinary speed was somewhere about four miles an
hour.

One hundred years ago news and letters travelled very
slowly, the post-boys to whom the letter bags were in-
trusted progressing at the rate of three and a half miles an
hour! In 1784 a proposal was laid before Government by
Mr. John Palmer, the originator of mail coaches, to run
quicker vehicles, though at much dearer rates of postage.
This scheme was at first opposed by Parliament, but after
a struggle of some two years, Palmer's coaches were adopted
for the conveyance of the mails, though the rate at which
these travelled was only six miles an hour for a long time
after their introduction.

A great impetus was given to the production of better
forms of stage coaches by gentlemen taking to drive them
as an amusement, and two clubs were soon formed of noble-
men and gentlemen who took an interest in four-in-hand
driving and in vehicles in general. Several clubs of this
kind are now flourishing to encourage manly sport, and
with the capacity to promote improvements in the form of
the "drag," as it is now called.

It is to an architect that we owe the invention of the
Hansom cab. The safety consisted in the arrangement of

the framework at the nearest part to the ground, so as to prevent an upset if the cab tilted up or down. The inventor was Mr. Hansom, the architect of the Birmingham Town Hall. Numberless improvements have been made on this idea, but the leading principles are the same.

In 1829 the first omnibus was started in London by Mr. Shillibeer, who some time previously had been a coach-maker in Paris. It was drawn by three horses, and carried twenty-two passengers, all inside. The fare was a shilling from the "Yorkshire Stingo," in Marylebone Road, to the Bank. This vehicle was found too large for the streets of London, so a smaller one was started, drawn by two horses and carrying twelve passengers inside. In 1849 an outside seat was added along the centre of the roof, and by 1857 the omnibus had become pretty nearly the same form as we now know it. Our present omnibus is probably the lightest vehicle of its kind for carrying such a large number of passengers. Its average weight is about 25 cwt. The London General Omnibus Company have, on an average, 626 omnibuses running on week-days, and 6,935 horses to work them. They build their own vehicles, and each runs about sixty miles a day, at a speed of about six miles an hour, and nearly all are supplied with brake retarders, worked by the foot, which effect a great saving in the strain put upon the horses in stopping.

CHAPTER II.

PREPARATION OF THE DESIGN AND SETTING OUT THE FULL-SIZED DRAUGHT.

In coach-building, as in building construction, the first thing to be done is to prepare a design of the vehicle proposed to be built according to the requirements of the customer. A scale of one inch to a foot is a very good one for the purpose, though the scale drawings are more often made to a scale of one and a half or two inches to a foot. These drawings (or draughts as they are technically termed), are prepared by specially trained draughtsmen, and it requires no mean skill to produce, on a small scale, a pictorial representation of the future vehicle, truly proportioned in all its parts, and a delicacy of touch in order that the parts may not look coarse. These drawings, if well made (and they generally are), give a very accurate picture of the carriage, and a purchaser is generally able from this to say what peculiar feature he requires, or where he thinks it should be altered; if he can do this it saves a great deal of trouble in the future, whilst the coach is being built.

For this work the draughtsman requires a drawing-board and T square, and two set squares; as he never has to prepare very large drawings, a board of imperial size will be amply sufficient, and the T square to have a corresponding length of blade. T squares are made of a variety of woods, but the most serviceable is one made of mahogany, with an ebony edge; the most important consideration being that the edge should be truly "shot" from end to end. The set squares should either be vulcanite or skeleton mahogany with ebony edges; the latter are preferable, as they work more cleanly than the vulcanite, which, unless kept very clean, are apt to make black smears across the drawing. In order to fasten the paper down to the drawing board, drawing pins will be required; they are simple pins of iron or steel, with a large flat brass head; four is the number required for each sheet of paper, one at each corner. A very much better way to fix the paper down is to "strain" it to the board. It is done in the following way :—The sheet of paper to be fastened down is thoroughly well wetted, by means of a sponge or large flat brush, on one side (which, it does not matter, but see that your board is perfectly clean before starting); it should then be left for five or ten minutes for the water to well soak into the pores of the paper; when this is done, the paper will be quite limp. Now take a perfectly clean straight edge, or the back edge of the T square, and turn up one of the edges of the paper $\frac{1}{4}$ or $\frac{1}{2}$ an inch against it ; along this edge run a brush charged with glue from the glue pot, or a piece of ordinary glue dipped into boiling water and rubbed along the edge will do just as well, and when you think there is enough sticky matter to promote adhesion between the paper and the board, turn the edge of the paper back on the board (without removing the straight edge or T square), and quickly rub it with the tips of the fingers until it goes down flat all along without any air bubbles: do this to all four edges of the paper, and place in a perfectly flat position to dry; and if the operation has been carefully conducted the paper will be beautifully flat to draw upon, and there can be no fear of its shifting. When the drawing is finished, all that has to be done is to cut round the edges of the paper just inside the glued edge, and take it off. A little hot water will take off the glued strip, and take care to wash all the

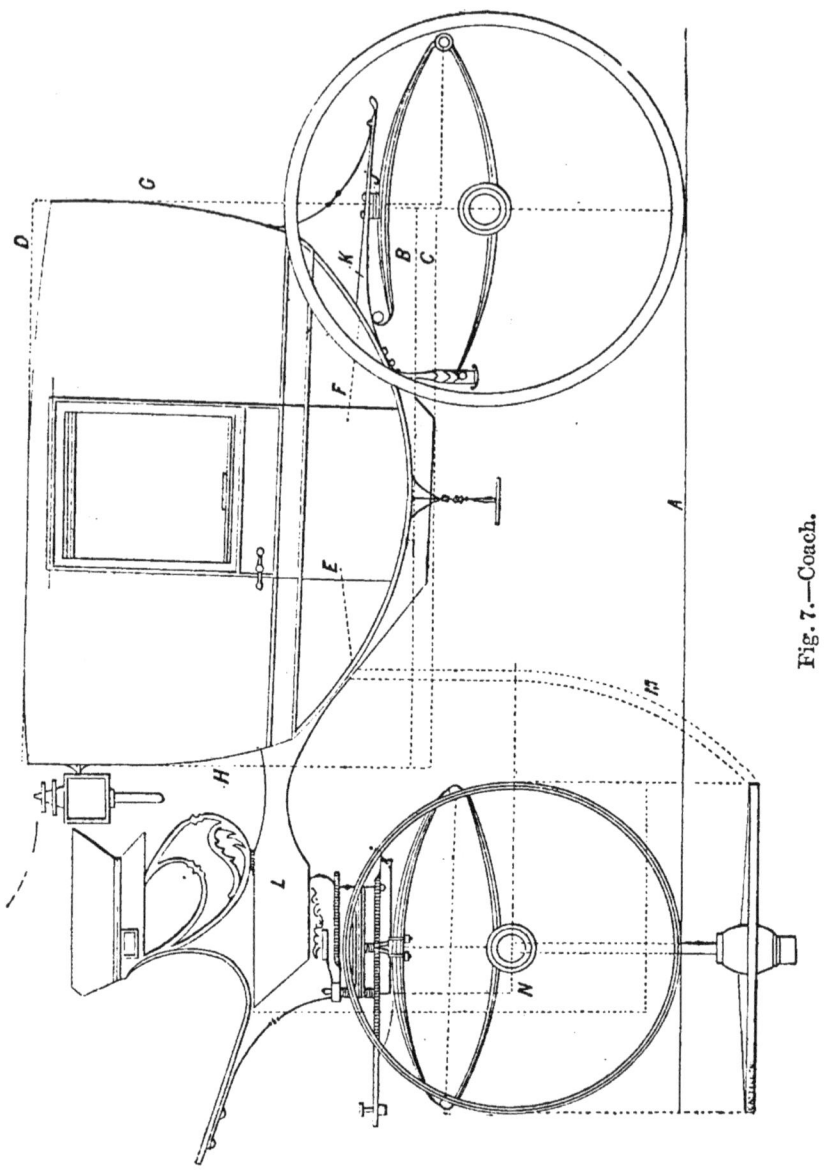

Fig. 7.—Coach.

glue off at the same time, otherwise a smaller piece of paper might stick in some important part, and the drawing spoilt in order to detach it.

The draughtsman will do well to have a few French curves, for drawing the " sweeps " or curved lines of the carriage bodies, and scales of various sizes, which are slips of boxwood or ivory, on which are marked at the edges various scales, from $\frac{1}{16}$th of an inch to a foot up to 3 inches to a foot ; and last, though not by any means least, a good box of compasses or mathematical instruments. We shall not discuss the merits of the various kinds of instruments here, but any one wishing to go into the matter may do so by reading " Mathematical Instruments " in Weale's Series. But we should strongly advise the draughtsman to go to some good maker, as bad drawing instruments only lead to bad drawing.

The drawing paper used should be of a kind having a slight gloss on the surface, like " hot-pressed " paper, but without its granular texture. This kind of paper is usually called a " board," as Bristol board, and kept in various sizes, and sold by all colour dealers. Various names are given to it, but it is all pretty nearly alike.

The paper being fastened, the drawing is commenced by drawing the ground line A (Fig. 7) ; from that set off the height that the body is to be from the ground, indicated by the dotted line B, and draw the line C, which is the depth of the rocker. This latter is the real bottom of the vehicle, and from it is measured the height of the seat, about 12 inches, shown by the dotted lines on the body. Then from the seat measure 42 inches, the length of the roof D. Lay off 23 inches for the width of the door, and draw E and F. From F measure 28 inches, the depth of the back quarter G, and from E measure 25 inches, which will give the front quarter H. Now the curves or sweeps of the body can be put in by means of French curves. From the hinge pillar measure 26 inches, shown by dotted line I, and this is the centre of the hind wheel, which is 4 feet 3 inches high. The spring is $1\frac{1}{4}$ inches thick, and consists of 5 plates 42 inches long. The opening between the springs is $12\frac{1}{2}$ inches, the lower one being clipped beneath the axle. Measure $12\frac{1}{2}$ inches from the underside of the axle, which will give the underside of the top spring. $1\frac{1}{4}$ inches must also be allowed for the back bar J, and the pump-handle K will be $\frac{1}{2}$ an inch thick. Then draw the boot L in such a position that the front wheels will lock or turn under it freely. This may be found by drawing a plan of the wheel as shown, and with the centre of the lock bolt produced to N, strike the lines M, and it will be seen that the wheels will just clear the body, which is all that is necessary. From this it will be noticed that the centre on which the fore carriage turns is not in the same plane as the axle. This is more particularly discussed in the chapter on wheel-plates. The front wheel is 42 inches high, the springs the same size as the back springs. The draught may be now completed from Fig. 7, after having settled on the various heights and sizes, and can be inked in with Indian ink. The dotted lines, being merely constructional, are rubbed out when the drawing is inked in. To complete the drawing, the spokes of the wheels must be shown. These should be neither too many nor too few, but there is no rule which regulates their number, except that there should be two to each felloe. Having inked the parts in and cleaned the pencil lines off, the drawing is ready to be coloured. The colours applied to the drawing are the same as will be used for painting the carriage, so we shall not detail them here.

From this drawing is constructed the full-size draught, which is prepared before a tool is touched. On the walls of the body-making shop are large black-boards, 10 or 12 feet square, and on these the draughts are prepared just in the same way as described for the scale drawing, except that all the heights are marked up a vertical line which runs through the centre of the doorway, and from this the various widths are also set off. This and the ground line are the first two lines drawn, and it is imperative that they should make a perfect right angle with each other, otherwise the draught will not be true, and the material worked from it will be wasted. This full-sized draught requires the greatest care in preparation, as all the patterns to which the materials are cut or shaped are taken from it, even to the smallest parts.

The full-size draught also differs from the scale draught, inasmuch as all the details of the construction of the vehicle are shown as in the accompanying cut (Fig. 8), which shows the construction of a small doctor's brougham, and Fig. 9, which shows the construc-

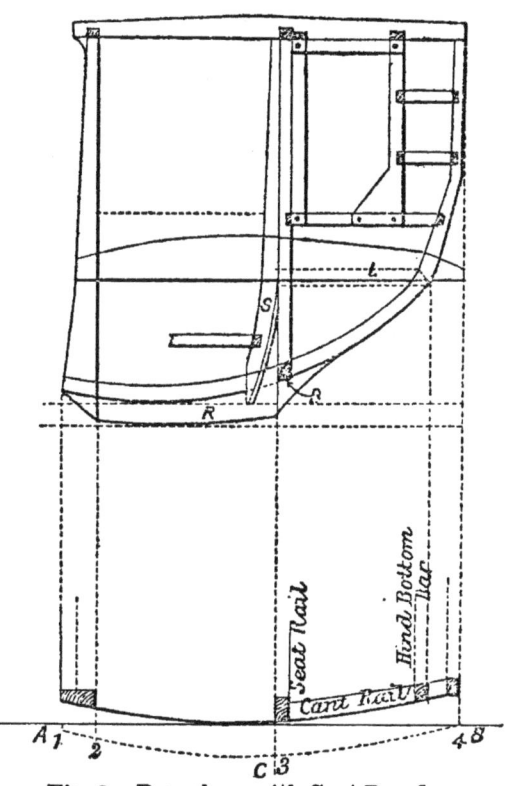

Fig. 8.—Brougham with Cant Board. s, Standing pillar (developed). B, Bottom bar. R, Rocker. L, Seat.

tion of a landau. This latter is a representation of the working draught for the vehicle, and, in fact, is a reduced copy of what would be drawn upon the black-board in the shop, except that some of the minor details are omitted to avoid confusion.

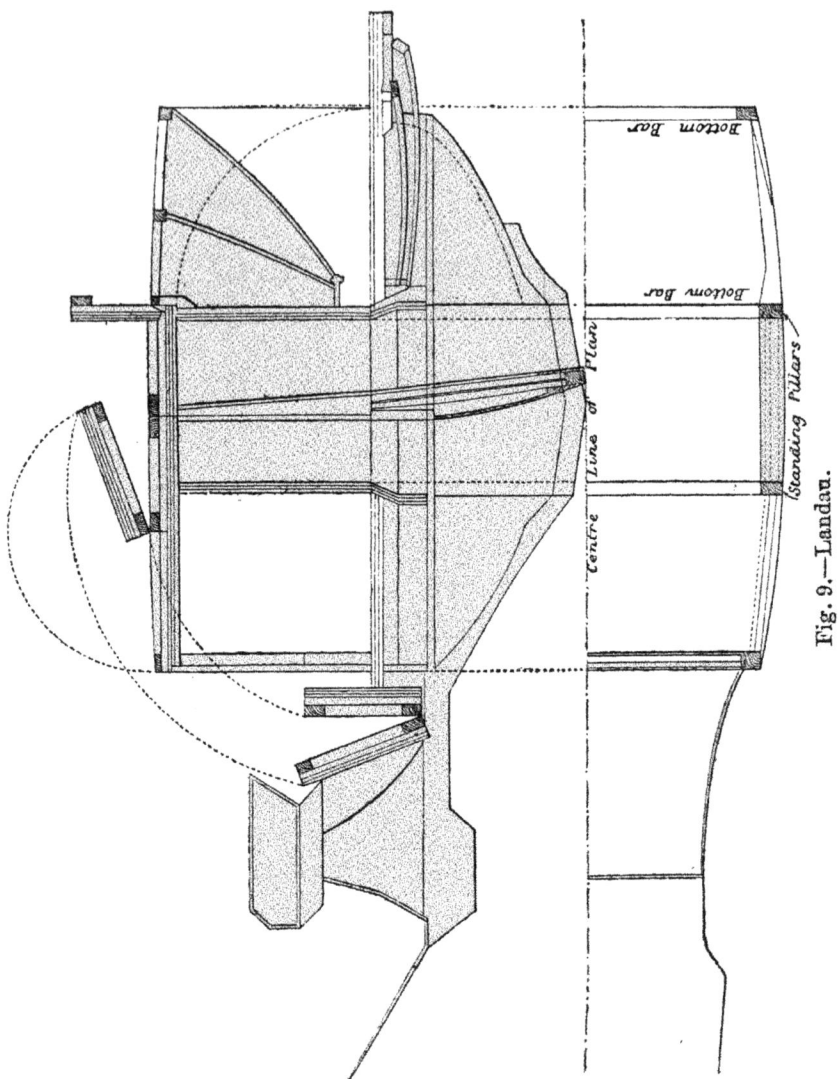

Fig. 9.—Landau.

CHAPTER III.

VARIOUS MATERIALS USED IN COACH-BUILDING.

THE materials employed in coach-building number a great many: various kinds of wood—ash, beech, elm, oak, mahogany, cedar, deal, pine, &c.; hides, skins, hair, wool, silk, glue, whalebone, ivory, &c.; iron, steel, copper, brass, lead, tin, glass, &c.

The timber principally used in the construction of carriages is the ash. This is not an elastic, but rather a tough and fibrous wood, capable of altering its form by the application of pressure, and therefore when not in large masses requires iron plates to secure it. By boiling it becomes very pliable, and may be formed into almost any shape, provided that it is not too thick. For this purpose it is better to use steam than boiling water, as the latter is likely to dissolve and carry off the gluten which unites the fibres, thus rendering the timber useless. Some ash timber is white at heart, and some red; the white is usually the strongest and best. Some trees which have been grown on hillsides much exposed to constant winds present a remarkably wrinkled appearance through their whole length, and it is scarcely possible to plane their timber smooth; this is the toughest of all ash timber. Parts of ash-trees are sometimes found of a yellowish-brown colour, accompanied by a fetid acid smell. This is sometimes attributed to the effect of lightning, but more probably it is a putrid fermentation of the sap, owing to imperfect drying. All other circumstances being equal, the timber is best which is cut down when the circulation of the sap is slowest, as the pores are then open. In the process of drying or seasoning the bulk diminishes considerably. One of the qualities which render ash peculiarly fit for carriage construction is the absence of elasticity, and consequent indisposition to alter its form by warping or twisting. It is not well adapted for boards or planks in which much width is required, as in drying it cracks a great deal. The diameter of ash-trees used by carriage-builders varies from 1 foot to 3 feet 6 inches. It should be borne in mind in cutting ash, that the interior and the outer casing under the bark are rather softer and less durable than the parts between them.

Beech is sometimes used by carriage-builders and by wheelwrights, on account of its cheapness; but it is very liable to warp and rot, and consequently unworthy of the attention of the conscientious manufacturer.

Elm is largely used for planking where strength is required. The grain is wavy, hard to work, brittle, and apt to split without care. It is not a good surface to paint on, as the grain shows through several coats of colour. It is also used for the naves or stocks of wheels.

Oak is used for the spokes of wheels. The best kinds are made from the timbers of saplings, which are not sawn but *cleft,* in order that the grain may be not cut across and render the spoke unfit to resist the strains it will be subject to. Spokes are also made from the limbs of large trees.

Mahogany is largely used for panels, as when painted it shows a very even surface. There are two kinds, the " Spanish " and the " Honduras." The former is unfit for the purposes of the carriage-builder. It is heavy and very difficult to work, requiring special tools for this purpose, as the edges of ordinary tools are rapidly destroyed by it. The Honduras is very much lighter and cheaper than Spanish, and the grain and colour more even. It takes the sweeps and curves required for body-work very easily. It can be procured up to 4 feet in width, straight-grained, and free from knots and blemishes.

A coarse-grained species of cedar is brought from the same district as Honduras mahogany, and is sometimes used for panels which have to be covered with leather, &c. Its extreme porosity renders it unfit for the application of paint.

Deal is largely used for the flooring of carriages, and for covered panels, and for any rough work that is not exposed to great wear and tear.

The wide American pine. is chiefly used in very thin boards to form the covered panels and roofing of carriages.

Lancewood is a straight-grained, elastic wood, but very brittle when its limit of elasticity is reached. It comes from the West Indies in taper poles about 20 feet long and 6 or 8 inches diameter at the largest end. It was formerly much used for shafts, but since curved forms have been fashionable it has fallen into disuse. It can be bent by boiling, but is a very unsafe material to trust to such an important office as the shafts.

American birch is a very valuable wood for flat boarding, as it can be procured up to 3 feet in width. It is of a perfectly homogeneous substance, free from rents, and with scarcely a perceptible pore. It works easily with the plane and yields a very smooth surface, and the grain does not show through the most delicate coat of paint. Its chief disadvantage is its brittleness, which will not permit of its being used for any but plane surfaces, and some care is required in nailing and screwing it.

Hides are used chiefly for coverings, but also in some parts strips are used for the purposes of suspension. The hides are those of horses and neat cattle. For covering they are converted into leather by the action of oak and other bark. They are afterwards smoothed and levelled by the currier, and sometimes split into two equal thicknesses by machinery. They are then rendered pliable by the action of oil and tallow, and finished to a clear black or brown colour as may be required. This is called dressed leather. For some purposes the hides are merely levelled, put on wet to the object they are intended to cover, and left to shrink and dry. Others are covered with a coat of elastic japan, which gives them a highly glazed surface, impermeable to water ; in this state they are called patent leather. In a more perfectly elastic mode of japanning, which will permit folding without cracking the surface, they are called enamelled leather. They are generally black, but any colour desired may be given to them. All this japanned leather has the japan annealed, somewhat in the same mode as glass. The hides are laid between blankets, and are subjected to the heat of an oven raised to the proper temperature during several hours.

The skins used are those of the sheep and goat. The former are converted into leather by the action of oak bark. In one form of dressing them they are known as basil leather, which is of a light brown colour and very soft. Sometimes they are blacked, and occasionally japanned like the hides. In all these forms sheep skins are only used for inferior purposes, as mere coverings, where no strength is required.

Goat skins are used in the preparation of the leather known as " Spanish " and " Morocco." They are not tanned in oak bark like other leather, but very slightly in the bark of the sumach-tree. They pass through many processes previous to that of dyeing, for which purpose they are sewn up with the grain outwards and blown out like a bladder. This is to prevent the dye from getting access to the flesh side. This beautiful leather was originally manu-factured by the Moors, who afterwards introduced the process into Spain, by which means it came to be known under two names. The English have greatly improved on the manufacture, so much so that few others can vie with it. These skins are used for the inside linings of carriages.

Hair is used as an article of stuffing. To give it the peculiar curl which renders it elastic, it is forcibly twisted up in small locks, and in that state baked in an oven to fix it. Horse-hair is the best, being the strongest and longest; but various other kinds are used. Sometimes it is adulterated with fibres of whalebone. Doe-hair is also much used as an article for stuffing, but as it is very short it cannot be curled, and there is not much elasticity in it.

Wool in its natural state is not used for carriage purposes. In the form of "flocks," which are the short combings and fibres produced in the process of manufacturing it, it is very largely used for stuffing. In its manufactured state wool is used in great quantities, as cloth, lace, fringe, carpeting, &c.

The iron used is that known as wrought iron. To judge of its quality break a piece over the anvil; if it breaks off brittle it is of no use for the purposes it is required for. If it is good wrought iron the fracture will present a bluish, fibrous, silky texture, without any crystalline portions. Inferior iron will either appear bright and glistening (when it partakes of the properties of cast iron) or dull and greyish in tone at the fracture.

It may also be tested by bringing it to a red heat and bending it, when any flaws, &c., will at once become apparent.

Cast iron is also used in the shape of axle-boxes.

Great quantities of wrought iron are used in the construction of modern carriages. One of the best qualities is that known as the "King and Queen," so called from its brand. This iron is manufactured from pieces of old iron, called scrap iron, which are placed in furnaces and welded under a heavy tilt-hammer, after which it is passed between rollers and converted into bars.

c 3

Steel also enters largely into carriage construction in the shape of springs, &c. Axles are made of Bessemer steel, and are found to wear very well. Steel consists of iron in which is combined a large proportion of carbon; the more carbon the higher the elasticity of the steel. If steel is over-heated, it gives up a portion of its carbon and approaches once again its original form of iron.

CHAPTER IV.

POINTS TO BE CONSIDERED BEFORE COMMENCING THE CONSTRUCTION OF A CARRIAGE.—COMPONENT PARTS OF THE BODY.—SMITH'S WORK.—GLUE.

As previously remarked the vehicle is divided into two parts —the carriage and the body. After the drawing or draught is carefully worked out to full size on the black-board in the shop, with all the curves and sweeps developed, and shown in elevation and plan, patterns or templates are made from the draught, and from these the construction of the body proceeds.

In commencing the construction of a vehicle there are several things to be borne in mind; such as the purpose to which the vehicle is to be applied, the size of horses to draw it, and other considerations arising from these two. It is popularly believed that the shorter the carriage the lighter it will run; in ascending an incline this may be true, but on ordinary level ground a long carriage and short one must be alike in friction, provided the total amount of weight and other circumstances be equally balanced.

Another consideration is the height of the wheels. On level ground, draught is easiest when the centre of the wheel is a little lower than the point of draught, viz. the point where the traces are affixed to the collar; but this in practice would be found rather inconvenient, as very high wheels would be required, and consequently the height of the whole vehicle would have to be increased, causing great trouble and annoyance in getting in and out of the vehicle, and the driver's seat would have to be raised to a corresponding height. Under equal circumstances a high wheel is more efficient than a low one, and requires less power to draw it; though it may be mentioned that a low wheel on a good and level road will do its work far better than a very much higher wheel on a rougher road. The sizes of the wheels of two-wheeled vehicles vary from 3 feet to 4 feet 6 inches.

It would be a very good thing if four-wheeled vehicles were to have the wheels of equal size, in order that the friction and power might be equal. But with the present mode of construction this is an impossibility, as we have only one mode of making the lock or turn. Therefore the height of the fore wheels must be regulated by the height at which the body hangs, so that the wheels may pass beneath it without striking, when the springs play. In practice this height varies from 2 feet to 3 feet 8 inches, according to the kind of carriage the wheels are intended for. The hind wheels vary from 3 feet to 4 feet 8 inches.

The next point is the dishing of the wheel, which is necessary for strength to take the strain off the nuts, to throw off the mud and prevent it clogging either the wheel or the body, and to give greater room for the body between the wheels without increasing the track on the ground. Whatever be the amount of dishing or coning, which varies from $1\frac{1}{2}$ to $2\frac{1}{2}$ inches, one rule should always be observed, viz. so to form the wheel that when running the lower spokes should maintain a true vertical position both in the fore and hind wheels. This is mainly accomplished by the dip of the axle, but if the fore and hind wheels have the same dish, they will take the same track along the ground. The dish of a wheel will be understood by referring to Fig. 10, in which it will be seen that the extremities of the spokes are not in the same plane, thus forming a dish or hollow in the surface of the wheel.

Some ingenious persons have deduced from the foregoing that a wheel runs best on an axle having a conical arm (the arm is the extremity of the axle which fits into an axle-box in the nave or stock of the wheel), in which case the axle would not dip, but the wheel would be put on to a perfectly horizontal axle. The motion of a wheel thus placed would be anything but artistic, though there would not be so much friction on an arm of this sort as on an arm of the dipped axle. Dipping the axle is shown at Fig. 10. It merely consists in bending it so far out of the horizontal as to give the lower spokes a vertical position. But in practice this theory of the conical arm will not answer, inasmuch as curving the

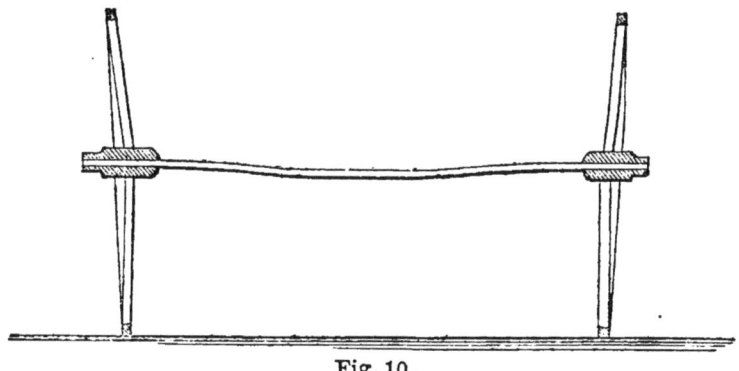

Fig. 10.

arm will reduce the front bearing surface so much that the oil would be squeezed out, and it would run dry, and the total amount of friction would be greatly increased. Long practice has shown that a cylindrical or slightly conical arm is the best that can be used.

We have now to settle the form, combination, and proportion of the springs. Springs which are laid on the axle at right angles have to carry the whole of the weight of the carriage, save only the wheels and axles. Where other springs are used in addition it is not necessary that the axle-springs should have much play. It will be sufficient to give them just so much play as will intercept the concussion caused by moving over a road. The strength of the springs must of course be adjusted to the weight they have to carry, for it is evident that if they be made sufficiently elastic to carry the weight of six persons, they will be found hard if only three enter the carriage. This is a disadvantage all carriages must labour under, for it is ridiculous to suppose that if a carriage is constructed to hold six that number will always want to use it at the same time. There would seem to be room for some improvement in the way of introducing springs adjustable to any weight, though, to give spring-makers their due, they do turn out really a first-class article in this respect; this is more noticeable because it is so recent. Light carriages are never so easy to ride as heavy ones, even when the springs are well adjusted, because on meeting with an obstacle there is not a sufficient resistance to the bound or jerk upwards of the spring, which makes riding in a light carriage over a rough road rather unpleasant.

The position of the front wheels next demands attention. As these have to turn under the body it requires some skill to fix them, and the play of the springs, the height of the axletree, and the height of the arch (the portion of the body under which they turn) have all to be considered. This will be more particularly described when dealing with wheel-plates.

The rule for the height of the splinter bar, to which the traces or shafts are fixed, is that it should fall on a line drawn from the horse's shoulder to the centre of the hind wheel. This, however, is not always convenient in practice, as the fore wheels regulate the height of the framing of the under carriage, to which the splinter bar is fixed. The distance of the splinter bar from the central pin, on which the wheel-plate and fore carriage turn, is regulated by the size of the wheels and the projection of the driving-seat foot-board.

All the above particulars are considered when setting out the full-sized draught, and all points capable of delineation are put on the board in some convenient part. In Fig. 9 the outline is simply given, as to show everything would only confuse the reader. Such other details as are required are filled in after the draught has reached the stage shown in the figure.

It is most necessary for the safe conduct of a coach and carriage builder's business that there should be a goodly stack of well-seasoned timber of the various kinds required, otherwise great trouble and vexation will arise in the course of business from a good piece of timber being perhaps spoilt in working, and there not being another piece in the factory to replace it.

Where there is sufficient accommodation it is usual for makers to season their own timber in specially constructed sheds, which are kept from bad weather, but at the same time thoroughly well ventilated. In these the timber is stacked, with small fillets between each plank or board, to insure a free current of air circulating all round. One year should be allowed for seasoning for every inch of thickness in the timber, and none should be used in which this rule has not been observed.

Thin portions of timber, such as panel stuff and the like, should be treated in the same way, and in addition the ends should be secured to prevent splitting. The panel stuff undergoes another process of seasoning after it is planed up ; in fact, all the thin timber required for roofs, sides, &c., does. And about the first thing done in commencing to build a carriage is for the body-maker to get his thin stuff ready, as far as planing it up goes, and then to put it aside in some moderately dry place, with slips of wood between each board to allow a circulation of air round them. The other stuff that is likely to be required should also be selected and put aside. If all these things be strictly attended to, there is not likely to be much trouble about bad joints; and it will be to the employer's interest to look after such workmen who have not enough scientific knowledge to see the reason of things themselves, and put them in the right direction. But an intelligent workman will soon appreciate the advantage of getting his stuff ready at the commencement, instead of waiting till he wants to use it.

The parts composing the body may be thus enumerated:—

The frame or case.

The doors.

The glasses, which are fixed in thin frames of wainscot, covered with cloth or velvet. It is a very good thing to have india-rubber for these to fall on, and little india-rubber buffers would prevent them from rattling.

The blinds, which are sometimes panel, but more generally Venetian, so adjusted with springs that the bars may stand open at any required angle.

The curtains, of silk, which slide up and down on spring rollers.

The lining and cushions, of cloth, silk, or morocco, as the case may be, ornamented with lace, &c. The cushions are sometimes made elastic with small spiral springs.

The steps, which are made to fold up and fit into recesses in the doors, or in the bottom, when they are not in use.

The lamps, which are fixed to the fore part of the body by means of iron stays.

The boot, on which is carried the coachman's seat.

In carriages suspended from C springs we have in addition:—

The check-brace rings, to which are attached leather braces from the spring heads, to prevent the body from swinging too much backwards and forwards.

The collar-brace rings, to which are attached leather braces from the perch, to prevent the body swinging too much upwards or sideways.

The curve or rounding given to the side of the body from end to end is called the *side-cant*, and the rounding from the top to the bottom the *turn-under*. Some makers arrive at this curve by framing the skeleton of the body together with square timber, and then round these off to the required curve *after* they are put together. It must be evident to any one that this proceeding will greatly strain the joints, and under any circumstances will never give thorough satisfaction or good results, and the waste of time and material must be very considerable.

The proper way is to set the curve out beforehand on a board called the " cant " board, and the method of doing this is as follows:—

Take a clean pine board, plane it up to a smooth surface. Shoot one edge perfectly true with a trying-plane. This straight edge may be taken to represent the side of the carriage if it were a straight line. Apply this edge to the full-sized draught, and mark along it the various parts of the body (see Fig. 8, in which the numbered points are those required to form the side-cant). By means of these points the required sweep can be set up or drawn, as shown by the dotted line c in the figure. Now, if you choose, you can cut away the portion between A and B, and a template will be formed to which the constructional timbers can be cut; and it possesses the advantage of being easily applied to the carriage as it proceeds, to see that the curve is true and uniform. As this template forms the pattern to which the timber, &c., is cut, great care is requisite in forming it, so that it shall be perfectly true.

In order to get the turn-under, the same process is gone through on another board. This gives what is called the " standing " pillar pattern, the standing pillar being the up-right timber to which the door is hinged.

There is no rule in particular for determining the amount of side-cant or turn-under to be given to a vehicle, 2½ or

3 inches on each side making the outside width of the body; 5 or 6 inches less at the bottom than at the elbow line is a usual allowance, but this is entirely dependent on the will or taste of the workman.

The cant-board described above is one having a " concave " surface; but it quite as often has a convex surface, and it is just as well to have one of each, and use the convex for cutting the timbers to, and the concave for trying them when in place, though, if this be done, it is imperative that the curves on the two boards should be one and the same. The same remarks apply to the standing pillar pattern.

The body is a species of box, fitted with doors and windows, and lined and wadded for the purpose of comfort. As the greatest amount of strain is put upon the bottom part, and the forces acting on the other parts are transmitted to the bottom, it is necessary that it should be very strongly put together. The two side bottom timbers are bonded, or tied together, by two cross timbers called bottom bars, which are firmly framed into them. To give depth to the floor, without destroying the symmetry of the side, deep pieces of elm plank are fixed to the inside of the side bottom pieces, and to these the flooring-boards are nailed, being additionally secured by iron strap plates, nailed or screwed beneath them. In the central portion of the bottom sides are framed the door-posts, called standing pillars.; At the angles of the bottom framework are scarfed the corner pillars. The cross framing pieces, which connect the pillars, are called rails. Two of these rails stretch across the body inside, on which the seats are formed; these are called seat rails. The doors are framed double, to contain a hollow space for the glasses and blinds, and they are fastened by means of a wedge lock, forced into a groove by a lever handle. There is a window in each door and one in front of an ordinary carriage, say a brougham. The doors are hinged with secret or flush hinges.

Before cutting the timber to the various sizes required, patterns or templates of all the parts are made in thin wood from the full-sized draught ; also of the various curves likely to be given to the different parts of the body.

Before a workman could be trusted with the making of a body, he must of course have considerably advanced in the knowledge of his craft beyond the mere use of his tools, because the success of a carriage depends very largely upon the individual skill of the workman, more so than perhaps in any other trade.

The stuff is marked out from the thin patterns before mentioned by means of chalk, and in doing so care should be taken to lay the patterns on the timber so that the grain may run as nearly as possible in a line with it, and thus obtaining the greatest possible strength in the wood, which lies in the direction of the grain. Thus if the pattern be straight, lay it down on a piece of straight-grained timber ; if the pattern sweep round, then get a piece of timber the grain of which will follow, or nearly follow, the line of pattern.

The strongest timber that can be obtained is necessary for the construction of the hind and front bottom sides ; for the weight is directly transmitted to these, more particularly the hind bottom sides, where the pump-handles are fixed.

The body-maker, having marked and cut out the various pieces of timber he will require, planes a flat side to each of them, from which all the other sides, whether plain or curved, are formed and finished. They are then framed and scarfed together, after which the various grooves are formed for the panels and rebates, for the floor-boards to fit on to. Then, if there is to be any carved or beaded work, it is performed by the carver. Previous to being fitted in, some of the panels have strong canvas glued firmly on their backs, and when fitted in blocks are glued round the internal angles to give greater security to the joints, and to fix the panels firmly in their places. Before the upper panels are put in, the roof is nailed on, and all the joints stuck over with glued blocks inside. The upper panels are then put on, united at the corners, and blocked inside.

If the foreman who superintends all this be a thoroughly skilful artisan, and the men under him possess equal intelligence and skill, the work might be distributed amongst almost as many men as there are parts in the framework of the body. These parts will be worked up, the mortises and tenons, the rabbets and tongues, being all cut to specified gauges; and when they are all ready it will be found that they go together like a Chinese puzzle.

The woodwork being completed, the currier now takes the body in hand, and a hide of undressed leather, specially prepared for it, is strained over the roof, the back, and the top quarters of the body whilst in a soft pulpy state, and carefully sleeked or flattened down till it is perfectly flat. This sleeking down is a rather tedious process, and takes a long time and a great amount of care to bring it to a successful issue; when it is flattened down satisfactorily, it is nailed round the edges and left to dry, which will take several days.

Such panels as require bending may be brought to the required sweep by wetting one side and subjecting the other to heat, as of a small furnace.

The doors are now made and hinged, and the hollow spaces intended to hold the glasses and blinds are covered in with thin boards, to prevent any foreign matter from getting down into the space, and being a source of trouble to dislodge.

In constructing the body the aid of the smith is called in. His services are required to strengthen the parts subjected to great strain, more particularly the timbers forming the construction of the lower portion. All along each side of the body should be plated with iron; this should be of the best brand and toughest quality. It is several inches wide, and varies from ¼ to ¾ of an inch in thickness. This is called the "edge plate," and is really the backbone of the body, for everything depends on its stability. It should run from one extremity to the other, commencing at the hind bottom bar, on to which it should be cranked, and ending at the front part of the front boot, bottom side. This plate should take a perfectly flat bearing at every point. Great care must be taken in fitting it, for although the plate may be of the requisite strength the absence of this perfect fitting will render it comparatively weak, the result of which will be found, when the carriage is completed and mounted on the wheels, by the springing of the sides, which will cause the pillars of the body to press on the doors, and it will be a matter of great difficulty to open them.

In the application of smith's work to coach-building, it is often necessary to fit the iron to intricate parts while it is red hot, and if due precaution be not taken the wood becomes charred and useless, and in cases where there are glued joints it may cause the loosening or breaking of these joints and other material defects. It is an easy matter to have the means at hand to get over the difficulty. All that is necessary is to have handy some heat neutraliser. One of the commonest things that can be used is chalk, and no smith's shop should ever be without it. If chalk is rubbed over the surface to which the hot iron is to be applied it will not char or burn. Plaster of Paris is a still more powerful heat neutraliser, and it is freer from grit. A small quantity of the plaster mixed with water, and worked up to the proper consistency, will be ready for use in about two hours. Many smiths will say that they never have any accidents in applying heated iron, but on inquiry the reason is apparent, for it will generally be found that such men use chalk, in order to see that the iron plate takes its proper bearings, thus inadvertently using a proper heat neutraliser. If it were more generally known that the difficulty could be met by such simple means, there would be less material spoilt in the smith's shop.

It has been very common of late years for body-makers to use glue instead of screws and nails for panel work, &c.; but it requires a great deal of experience for a man to use glue with successful results. It is useless for the tyro to try it; he will only spoil the work. So, unless the artisan be well experienced in the treatment and application of glue, he had better leave it alone. To render the operation successful two considerations must be taken into account. *First:* To do good gluing requires that the timber should be well seasoned and the work well fitted. *Second:* In preparing for gluing use a scratch plane or rasp to form a rough surface of the pieces to be joined together, for the same purpose that a plasterer scores over his first coat of plasterwork, in order to give a key or hold. The shop in which the gluing is done should be at a pretty good temperature, and so should the material, so that the glue may flow freely. Having the glue properly prepared, spread it upon the parts, so as to fill up the pores and grain of the wood, and put the pieces together; then keep the joints tight by means of iron cramps where it is possible, and if this cannot be done the joints must be pushed tightly up, and held till the glue is a little set and there is no fear of its giving way. All superfluous glue will be forced out by this pressure and can be cleaned off.

A great cause of bad gluing is using inferior glue and laying it on too thick. Before using a new quality of glue, the body-maker should always test it by taking, say a piece of poplar and a piece of ash, and glue them together, and if when dry the joints give way under leverage caused by the insertion of the chisel, the glue is not fit for the purposes of carriage-building and should be rejected. With good glue, like good cement, the material should rather give way than the substance promoting adhesion. This is a very severe test, but in putting it into practice you will be repaid by the stability of your work.

Waterproof Glue.

It is often found that joints glued together will allow water to dissolve the glue, and thereby destroy its adhesive power. It may have been well painted and every care taken to make it impervious to water, but owing to its exposed position water has managed to get in. Often where screws are put in the glue around them will be dissolved, caused by the screws sweating; and it is very often found, where the screws are inserted in a panel, that the glue loses its strength and allows the joint to open, and there is little or no appearance of glue on the wood, which shows that it has been absorbed by the moisture.

To render ordinary glue insoluble, the water with which it is mixed should have a little bichromate of potash dissolved in it. Chromic acid has the property of rendering glue or gelatine insoluble. And, as the operation of heating the glue-pot is conducted in the light, no special exposure of the pieces joined is necessary.

Glue prepared in this manner is preferable in gluing the panels on bodies, which are liable to the action of water or damp. The strength of the glue is not affected by the addition of the potash.

In plugging screw holes glue the edge of the plug; put no glue into the hole. By this means the surplus glue is left on the surface, and if the plug does not hit the screw it will seldom show.

Where brads are used the heads should be well set in; then pass a sponge well saturated with hot water over them, filling the holes with water. This brings the wood more to its natural position, and it closes by degrees over the brad heads. The brad must have a chance to expand, when exposed to the heat of the sun, without hitting the putty stopping; if it does it will force the putty out so as to show, by disturbing the surface, after the work is finished.

CHAPTER V.

WE have now to consider the construction of the lower framework, or *carriage*.

The following is a list of the chief parts of a *coach*, as generally known :—

Wheels.

Axles.

Springs.

Beds, or cross framing timbers, which are technically termed the fore axle bed, the hind axle bed, fore spring bed or transom, hind spring bed, and horn bar.

Perch, or central longitudinal timber connecting the axletrees.

Wings, which are spreading sides, hooped to the perch and framed to the hind beds.

Nunters, or small framing pieces, which help to bind the hind beds together.

Hooping-piece. A piece of timber scarped and hooped to the fore end of the perch to secure it to the

Wheel plate, which is the circular iron beneath which the fore carriage turns.

The fore carriage consists of the fore axle beds, into which are framed the

Futchells (French, *fourchil*, a fork), which are the longitudinal timbers supporting the

Splinter-bar and the

Pole, to which the horses are attached.

The hinder ends of the futchells support the

Sway-bar—a circular piece of timber working beneath
the wheel plate.

A circular piece of timber of smaller size, supported on the
fore part of the futchells for a similar purpose, is called the

Felloe-piece (often made of iron).

On the splinter-bar are fixed the

Roller bolts, for fastening the traces.

On the pole is fixed the

Pole hook, to secure the harness.

The perch and beds are strengthened with iron plates, where
necessary, and the other ironwork consists of

Splinter-bar stays, to resist the action of the draught.
Formerly these were affixed to the ends of the
axles and called " wheel-irons."

Tread-steps, for the coachman to mount by.

Footman's step.

Spring-stays.

On the beds are placed

Blocks, to support the

C springs ; to which are attached

Jacks, or small windlasses, and

Leathern suspension braces.

These parts fitted together would form what is generally
known as a coach, or a vehicle, the body of which is large,
and suspended by leathern braces from the ends of C springs.
They enter into the formation of all vehicles more or less,
but for the other kinds some part or parts are omitted, as in
a brougham hung on elliptic springs, the C springs, perch,
leather braces, &c., would be omitted, and, of course, elliptic
springs and a pump-handle would be added. All the wood-

work is lightened as much as possible by the introduction of beading, carving, chamfering, &c.

In starting the carriage part the workman first takes the perch and planes a flat side to it, and then works it taper from front to back. The top and bottom curves are then worked up, or at least some portion of them, and then the front and hind spring beds are framed on. A pair of spreading wings are then fitted to the sides of the perch; these are simply circular iron stays, swelled and moulded to take off their plainness. A pair is fitted at each end of the perch. The hind axletree bed is then scarfed upon the top of the perch and wings, and is connected with the hind spring bed by two small framing pieces called nunters. At the front end of the perch a cross bed called a horn-bar is scarfed on the perch, at the same distance from the fore spring bed as the hind axle bed is from the hind spring bed, viz. the length of the bearing of the spring, or about 15 inches. The horn-bar is connected with the fore spring bed by the two spring blocks, which are either framed into them or scarfed down on them, and also by the hooping-piece, which is scarfed on the top of the perch. The perch is then planed up to the curve it is to have when finished, and it is then taken to the smith, who fits and rivets on the side plates, which have ears at the ends for the purpose of bolting them to the beds. The carver then does his work by beading the perch and beds, having due regard to the finish of the parts, rounds and curves all the ends. On the under side of the perch is riveted an iron plate, and on this plate is an iron hook for hanging the drag shoe and chain (if such be used). The hind framing is now put together, all connections being by means of mortises and tenons secured by screw bolts. The wings used to be, and sometimes still are, of wood, in which case they are hooped to the perch by iron hoops, and are rebated to receive the perch plates. The hooping-piece is then hooped in a similar way to the fore end of the perch, and the transom firmly bolted. The carriage is then turned bottom upwards, and the smith fits to the fore part the wheel plate or turning iron, across which runs a broad plate the width and length of the fore spring bed. A similar plate runs across the hind spring bed. The hind axle is then fitted to the wings and perch, and let into its bed at the ends, where screw clips secure it, the bolts passing through the perch.

The carriage is then turned up into its old position; the wheel-plate is cased on the top with carved wood, and a plate is riveted to the side of the horn bar. The springs are now fitted to their blocks and bolted firmly down. Iron stays are bolted to the springs beneath the beds to render them still firmer. The footman's step, and the steps for the coachman to mount to his box by, and other ironwork that may be required in the shape of stays, &c., are then fixed in thetr place.

The under portion of the fore carriage is framed to the fore axletree bed, which is a very stout piece of timber. Through this are framed the two futchells which receive the pole. The upper part of the axletree bed is covered with a strong plate to match the wheel plate. A circular piece of timber, called a sway-bar, is bolted behind the axletree bed, and this is also plated beneath for security. In front is a smaller piece of the same kind, and they both serve for the circumference of the wheel to rest on. The splinterbar is bolted on to the fore end of the futchells and secured by branching stays, one at either end connecting it with the axletree bed. As an additional security, iron stays are fitted to the bottoms of the futchells passing over the axle, which, in addition to bolts, is secured by screw clips at the ends, the same as the hind one.

The carriage above described is one suspended only on C springs. Sometimes elliptic springs are used in conjunction with C springs, and the former are then termed under-springs. In the latter case, of the double combination of springs, the constructional timbers may be of a less size or scantling, owing to some of the strain and concussion being removed. In this case the axles are clipped to the under-springs; but the general mode of construction is the same.

In first-class work a wrought-iron perch is used instead of the before described wooden one. This generally follows the contour of the underside of the body, and is called a swan neck. It enables the perch to be constructed of a much lighter appearance, and being really light, and to a certain extent elastic, all the beds and iron stays may be proportionately reduced in weight. The wheels and axles also, having less to carry, may also be made lighter. The system was introduced by Messrs. Hooper about 1846, and at first was only applied to broughams and sociables, but it has gradually been applied to the largest carriages, especially barouches and landaus. These perches are supported on horizontal under-springs, and are not now made so light as at first, for it is found that unless the hind wheels follow steadily, not only is the carriage heavier behind the horse, but the perch itself is frequently bent against very small obstructions; a stronger and stiffer perch is therefore now used, and it is found easier both to the horse and to the passengers.

When the body is suspended from C springs by leather braces, great care should be used in the selection of the material for these latter, and for this purpose the best and strongest leather is required.

The use of brake retarders to the hind wheels has now for some years superseded the old-fashioned drag shoes. It is evident that the action on two wheels must be better than on one only. The brake can be applied or removed without stopping the carriage, which is necessary if a drag shoe be used. This is rather an important consideration in undulating country, for it would be a great inconvenience to have to get down and put on the drag shoe when descending a hill, and when at the bottom to stop and get down again to remove it, in order to proceed along level ground or up the next hill, and so keep on like this all day. The lever brake was the original form, as still seen on drags, &c., but in many parts it is superseded by the foot or treadle brake, more especially in Scotland. This kind of brake is also the one used by the London Omnibus Company. The blocks which press upon the wheels have been made of various substances—cast iron, wrought iron, brass, wood, india-rubber, and leather. The wood is the best for the hold on the iron tyre and absence of noise and smell, but it wears out fast. India-rubber, especially for light carriages, seems to be the most satisfactory.

We have given, generally, the operation of framing together the under or carriage parts of the vehicle. But as some very important considerations regulate the shape, con-struction, and formation of most of these parts, they must be discussed separately. For this purpose they will be con-sidered under the following headings :—

Wheels.

Axles.

Springs.

Wheel-plates and fore carriages.

Ironwork generally.

CHAPTER VI.

WHEELS.

A WHEEL for a locomotive vehicle is a circular roller, either cylindrical or conical, the width or thickness of which is considerably less than its diameter. It may be either solid or constructed of various pieces, in which latter case it is called a framed wheel. It may also be made of wood or metal, or a combination of both.

Wheels which were made before the introduction of iron were of course very clumsy in their construction, in order to obtain the requisite strength. Specimens may still be seen in the broad wheels of waggons, technically termed rollers. The naves of these wheels are of enormous size. But when the naves of wheels were reduced for the purposes of elegance, a thin hoop of iron was applied both to the front and back, to prevent them from bursting by the strain on the spokes. When the felloes of wheels were reduced in size, straps of iron, called strakes or streaks, were applied to their convex surfaces covering the joints. But the last improvement was the most important of all, namely, the application of a " hoop-tire " instead of what was called the " strake-tire." Mr. Felton, in his treatise on coach-building, 1709, says, " Many persons prefer the common sort of wheel on account of their being more easily repaired than the hoop-tyre wheel; but though repairing the latter is more difficult, they are much less subject to need it."

The earliest form of wheel was no doubt a slice of the

trunk of a tree; portions of this being cut out for the purpose of lightening it would be the forerunner of spokes, or we should think the pieces left running from the nave to the felloe would be. The only improvement then effected for a very long period was making them of different pieces of timber instead of all from one piece, though of course the proportions of the parts would be considerably improved, if only for the sake of appearance.

At the end of the seventeenth century, among the wealthier classes, decoration was applied to coaches generally, and wheels in particular, to an extent which would surprise us nowadays. These latter were again ornamented as in the times of the old Roman Empire; the spokes were shaped and carved, the rim moulded, and the naves highly embossed; though, as may be imagined, there was a great want of taste in the application of all this ornament.

Towards the end of the eighteenth century, the extreme height of wheels extended to 5 feet 8 inches, which had but 14 spokes; wheels 5 feet 4 inches high had 12 spokes; wheels 4 feet 6 inches had 10 spokes; and the lowest wheels, 3 feet 2 inches high, had 8 spokes. The naves were of elm, the spokes of oak, and the rims or felloes of ash or beech. The rims of the higher wheels were often of bent timber, in two or more pieces, and were bolted to the tires by one bolt between each pair of spokes. The tire was put on in pieces, until the hoop tire came into general use, when it superseded the old ones entirely. In consequence of the great height of the wheels it was necessary to make the carriages very long, and the distance from the front to the hind axletree was 9 feet 2 inches in a chariot, and 9 feet 8 inches in a coach, or about 8 inches longer than we should consider necessary now.

These extreme sizes are now very seldom used, except in the case of large dress or state carriages and coaches.

The form of wheel now generally preferred in practice is

of the dished or conical kind, and the axle-arm on which it revolves is sloped or bent so far out of the horizontal that the lower spokes are in a vertical position. Undoubtedly the friction is increased by this arrangement, because a wheel on a horizontal axle runs easiest and smoothest; and when the axle-arm is slanted downwards towards the point the wheel has a tendency to bind harder against the shoulder which butts against the nave of the wheel, and the friction between the two is greater than would be the case if the axle-arm were perfectly horizontal. This, however, is a very small objection, inasmuch as this collar is firm and strong, and well fitted to bear any strain that may be thrown on it by the wheel; whereas, if the force acted in the other direction, or against the nuts and linch-pins, there are very few that would last out a day's journey. The advantage of throwing the strain on the firm and strong shoulder, which is well able to withstand it, is evident; and in this case, in the event of the nuts or linch-pins falling off or giving way, there is not so much danger of the wheel coming off at the moment the nuts go, as its tendency when on a level road is to run upwards towards the shoulder. Besides this, as the lower spokes are in a vertical position, the upper ones spread considerably outwards, and thus afford a greater space for the body between the wheels without the track on the ground being increased; and another advantage is that the mud collected by these conical wheels is thrown off away from the carriage.

The hind wheels of an ordinary carriage vary from 4 feet 3 inches to 4 feet 8 inches; the fore wheels are from 3 feet 4 inches to 3 feet 8 inches. The number of felloes in each circumference varies according to the number of spokes, two spokes being driven into each felloe; 14 to 20 spokes are a usual number for a hind wheel, 12 to 18 for a fore wheel; however, there is no rule to guide one in the matter, experience being the only teacher.

The wheels should be made with a due regard to the offices they have to fulfil; but we are inclined to think that this branch of the trade has not received that careful study which it deserves. Coach-makers seem in such a hurry to produce a perfect vehicle all at once, instead of beginning by improving the parts and then applying these improvements to the whole.

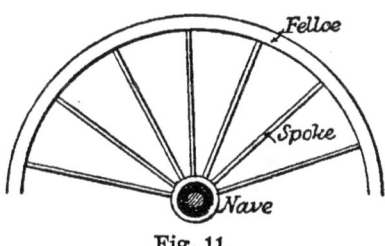

Fig. 11.

The mode of constructing a wheel is as follows :—

The timbers that are to be used should be well and carefully selected. The nave or stock, which is sliced from the limb of a tree, should be as nearly as possible the size required in its natural growth, so that it will require little reduction beyond what it receives in the lathe in bringing it up to the true circular form. The reason of this is that the annual rings which mark the grain of the timber should be as little disturbed as possible, as they are not all of equal strength and durability, the outer rings being pretty strong, but as they get nearer to the centre the wood is much softer. If, then, this outer hard casing is cut away, even only in part, it is signing the death-warrant of the poor nave, for the

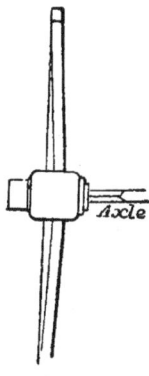

Fig. 12.

interior parts of the timber are not nearly so capable of resisting the destructive influences around, and in a very short time they will become completely soft and rotten.

As already remarked, the spokes should be cleft, not cut. The felloes which form the outer periphery of the wheel should also be cut as closely following the grain as possible.

When the wheelwright has carefully selected his timbers, he commences work by turning the stock in the lathe to the size required. Then he marks with a gauge of the same width as the spokes 4 circles as shown at *a a a a*, Fig. 13. The first and third of these mark the position of the front or

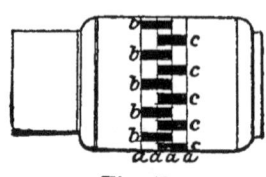

Fig. 13.

face spoke, and the second and fourth mark the position of the back spoke. Two holes are then bored in each mortise in succession, after which they are squared out with proper chisels. Truth of eye and skill of hand are the workmen's only guide in this operation, though it is evident that it is the most important operation of the whole, as upon it depend the accuracy and solidity of the wheel when finished. The tenons of the spokes are then cut to fit the mortises, parallel in their thickness, but in width they are cut slightly taper-wise, *i.e.* the extremity of the tenon is made about the same size as the mortise, but at the shoulder it is about one sixteenth of an inch larger, so as to make sure of the tenon filling the mortise when driven home.

Before cutting the mortises the stock should be fixed at some convenient angle, regulated by the amount of dish it is intended to give the wheel. This is particularly necessary, or when the felloes come to be fitted, if the mortise-cutting has been done in a slovenly way, the dish will not exist at all, or if it does it will be in the wrong direction.

Each alternate spoke is now driven in by the blows of a mallet to a perfectly close bearing of the shoulder of the

Section of Spoke

Fig. 14.

tenon, the workman guiding it as best he can. But it is evident that the position that the spokes will take is by no means certain. Owing to the wedge-like form given to the tenon, the spokes are driven home very tight, and wood not being of a homogeneous texture will yield more in one part than in another; and the mortise, cut in the way that it is, must be to some extent uncertain.

Every alternate spoke being driven, or those in the same plane, the remainder are driven in between them in the same manner.

In Fig. 13, *b, b, b,* are the mortises for the face spoke, and *c, c, c,* the mortises for the back spoke. Fig. 14 shows a section of an ordinary spoke, the hatched part showing the form of the greater part of its length, and the plain lines completing the rectangle show the extent of the swelling at the shoulder of the tenon at the nave.

Fig. 15 shows a very handy adjunct to the wheelwright's shop; it is called the "centring square." It is found extremely useful in marking and setting out the mortises for the spokes. Its construction is very simple, being but a T square, whose stock is the segment of a circle. A is the blade, B the circular stock, the extremities of which should

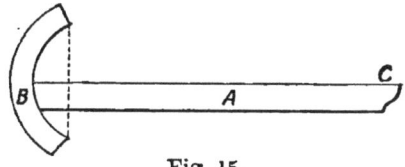

Fig. 15.

be protected by steel or brass, or, better still, have a steel edge round the whole of the inner surface of the stock, so that it will always keep true; for if it wears at all, of course its true circular form will be destroyed, and it will be rendered useless. And another important thing should be borne in mind, and that is to make the upper edge, c, of the blade in a line with the centre of the curve from which the circular head is struck; the reason of this will be apparent to the most obtuse, for unless the lines radiate from the given centre it is useless for the wheelwright's purpose.

After the spokes have been driven in they are shaved off by the spokeshave to their proper form, Fig. 14; and the lengths being measured from the nave, the outer tenons are cut, sometimes square, sometimes cylindrical, but leaving the back shoulder square to abut on the felloe with greater firmness. In the manner of tonguing there is a great deal of difference of opinion amongst wheelwrights; that the tongues in size should be slightly in excess of the hole or mortise to receive them, is a generally received idea, but a difference of opinion exists as to the length. But certainly it seems more reasonable to cut the tongues a little *shorter* than the length of the hole in the felloe to receive them, because when the tyre is put on it shrinks in cooling and draws up all the joints. Now suppose the tongues are a little longer than the holes, the shrinking of the tire causes it to press on these; and as they are firmly fixed at the nave, there is no escape for them, and the result is that the spoke is seriously crippled. It is fair to say that there are many good practical men who work in this latter way, and with to all appearance good results, but it is evident that the principle is not a good one.

One of the difficulties in making light carriage wheels is to get the spoke tightly into the felloe without splitting it, and the manner of accomplishing this is more successfully done, not by making the tongue on the spoke so large that it will fill the hole in the felloe of itself, but by making the tongue rather smaller and slitting the edges after it has been driven in, and then wedging up with small wedges, just in the same manner that a joiner would wedge up the mortises and tenons of a door.

When the wheel is so far progressed with it is laid on the ground, and the felloes are ranged round it in the order they are to be fitted. It is of the greatest importance that the holes should be bored in the felloe in an exact radial line from the nave; if this is not done, the spoke will have to be strained out of the straight line in order to get it into the hole; this will put an undue pressure upon it, and it is very likely that before the wheel has been long in use the spoke will break off short at the felloe. The exact position of all the mortises and joints should, therefore, be worked out on a full-sized drawing, and this being accurately done, there will not be much danger of going wrong. As the construction of a wheel is somewhat analogous to the arch, it is considered that by giving the felloes a number of joints the strength of the wheel is very much increased. Whether this be so or not must be left to the theorists to determine, as we have no trustworthy results from the various experiments under this heading.

The number of felloes in a wheel is decided by its size and number of spokes, two spokes being driven into each felloe. For an ordinary-sized brougham the felloes should be seven in number for the hind wheels, and six for the front wheels, or fourteen and twelve spokes respectively. For the purpose of connecting the felloes, a dowel or pin is cut on the end of one of them, and a corresponding hole bored on another, and they are fitted together; in common work holes are bored in each felloe, and an independent pin of hard wood or iron fitted into them. There is less time and labour consumed in this latter method; but the felloes constructed on this plan are not very reliable, and their weakness is soon shown by what is known as " dropping," which is simply caused by the wear and tear to which wheels are subject working- the dowels in the holes and enlarging them (the holes), and destroying the truth of the joint, which loss is soon discovered by the play or freedom given to the felloes allowing them to slip out of their place. But it must be borne in mind that this defect is just as liable to take place in felloes put together in the other way if the holes are not truly bored, and the joints are not well fitted.

The following directions as to putting on the tires are given in the " Coachmaker's Handbook," an American work :—

" First examine the wheels and see what condition they are in for the tire, so that we can determine what draught to give them. See if the felloes are drawn snug on the shoulder of the spokes, and how much open there is in the rim ; for instance, one set we will suppose to be $1\frac{1}{2}$-inch felloe, open $\frac{3}{16}$ of an inch, give a good $\frac{1}{4}$-inch draught; $1\frac{3}{8}$-inch felloe, $\frac{3}{16}$ open, just $\frac{1}{4}$-inch draught; $1\frac{1}{4}$-inch felloe, $\frac{3}{16}$ open, $\frac{3}{16}$ draught, and so on.

"Now in determining what draught to give the above wheels, we supposed them all to be good, sound, hard, hickory felloes; if the felloes are of soft timber just give them a trifle more draught. If the wheels should be above $\frac{1}{4}$-inch dish, the felloes would want only one-half the opening, but give them the same draught as the above. In running the tire, lay all the above tire in sets on the floor, roll the wheels on them, and allow 1 inch for taking up in bending; then mark the end of the nave with chalk, 1, 2, 3, 4, &c., and the tire with a sharp cold chisel mark I., II., III., IV., &c. Then straighten on a block set up endwise about 2 feet high, a little concave or hollow in the centre, letting the helper strike while the smith manages the tires until the kinks are all taken out of them. Then bend one end a little, so that it can be got into the machine, and take pains to get them as round as possible.

"In running the wheels with a 'traveller,' a wedge must be driven in one of the joints of the felloe for the purpose of tightening the other joints in the rim. Then get the length of the felloe, and in running the tire cut it $\frac{1}{8}$ inch shorter than the rim measures. In this explanation we are supposed to have steel tire, and we have a kind of steel tire now that is very high and difficult to weld, and there are many smiths that will profit by this lesson if they attend to the precaution we give. This tire steel will not stand as heavy heat as even cast steel, and if it is over-heated in the least it will crack or break in two while hot.

"There is one peculiar fact connected with it that we find in no other kind of steel, and that is this: it is apt to slip, however good the heat, and to obviate this, after scaffing the ends down to a sharp edge, make a rather sharp lap, and while hot take a sharp-pointed punch and punch a hole nearly through both laps, and drive in a sharp pin made of

$\frac{3}{16}$-inch steel wire and $\frac{1}{2}$ inch long. This will not show on the outside of the tire when on the wheel, neither does it weaken the tire like a rivet. We have often seen tires broken where the rivet went through.

" In welding, first have your fire perfectly clean, the coal pretty well charred, and the fire hot, but rather small, for the smaller the fire, if hot, the less it will waste your tire each side of the weld ; have the borax charred ; put a little on the weld while hot, pull the fire open with the poker and place the lap in the hottest part ; roll a few pieces of coked coal on the weld, blow steadily, carrying your tire back and forward through the fire, or stop the blast a moment until the lap is heated alike all through ; take it out and weld with hammer and sledge. With this precaution you will never fail getting a good welding heat, and need not upset your tire before welding. If your tire is upset before welding it makes the lap so much thicker that before it can be heated through alike there is a liability to over-heat, and waste away the tire on each side of the weld.

" In laying the tires down the heaviest should be laid at the bottom, and levelled with brick, so that the tire will rest permanently on every brick or bearing, and the rest laid on top the way they will fit best, to prevent warping the tire in the fire. A level stone should be used to lay the wheel on when the tire is put on. If the tires do not get warped in the fire do not hammer them at all, without there are some kinks left in the tires in fitting them ; avoid hammering if possible, for it marks the tire ; cool off, gradually pouring the water on out of the spout of a tea-kettle until it shrinks enough so it can be taken up ; then roll it in soap-water to prevent it from hardening, until it is so cool that it will not burn the felloes, truing up while the helper is rolling it in the water with a mallet covered with thick leather at both ends ; let the third person take the wheel and finish truing the tire with a leather-covered mallet ; while it is so hot that you cannot bear your hand on it the felloes move easily under the tire, and it should not be moved after it has cooled off if it can be otherwise avoided, for this reason, when the tire gets cold all its roughness and imperfections become embedded in the felloe."

" The tire once moved will move the easier next time. After the tires are all on examine the wheels, and see if there are any crooked spots in the tire that do not set down to the rim ; should there be any, heat a short piece of iron and lay on the tire, it will soon heat it enough to burn the felloe, but take it off before that time, and rap it down with a hammer. It is a bad practice to heat the tires on a forge as some do, for in truing them in fitting we have to bend them cold, and if heated on the forge and one place red hot, you will often find there a short crook edgewise. If some of the wheels are dished more than the others, put them on the off-side of the carriage. Never take a tire off if it can be avoided, without it is so loose or tight as to spoil the wheel when run."

When the tire has got sufficiently cold it is riveted to the felloe by countersunk rivets, one on each side of the felloe joints.

The strength or weakness of wheels plays an important part in the durability of the vehicle, for in whatever manner the various forces are mechanically met they at last concentrate themselves on the wheel ; it is highly necessary, therefore, that great pains should be taken in constructing them. The stock is not necessarily the foundation on which to build a wheel, and, further, there are many objections to its being so constituted. In the first place, when its centre is all scooped out for the reception of the axle-box, and its sides are mortised out to receive the ends of the spokes, it is nothing but a mere shell. Every mortise hole is more or less a receptacle for water, which the best workmanship cannot wholly exclude ; and as one part of the stock is always more porous than another, that is the part that will soonest absorb wet and begin to decay. If, therefore, the stock could be dispensed with greater durability would be insured.

A thoughtful inventor, turning over these things in his mind, has, during the last few years, produced a wheel of novel construction, which is found practically to be superior to the one in common use. All the spokes, instead of being shouldered down to enter the stock, are made wedge-shaped at the end, and instead of the wheel being constructed from the centre to the felloe, it is constructed from the felloe to the centre. Every felloe is made and fitted with its two

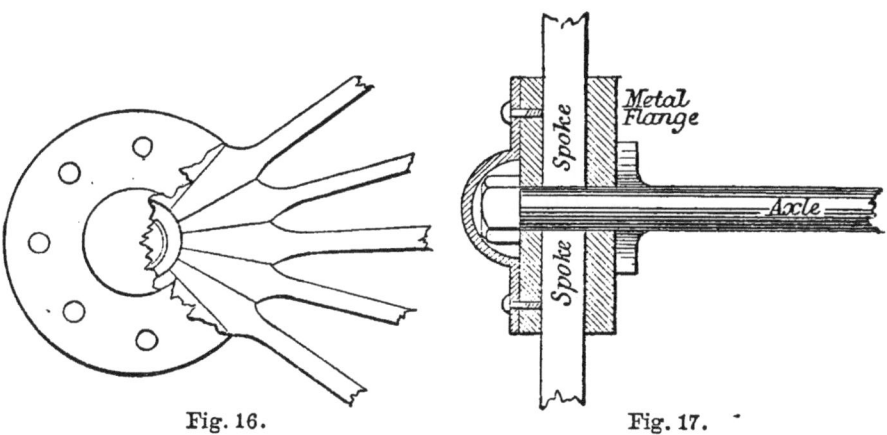

Fig. 16.   Fig. 17.

spokes, which, as they converge towards the centre, press upon each other in such a manner that when the whole periphery is put together a solid centre is produced by the spokes themselves, as shown in Figs. 16 and 17 ; so that instead of being dependent on wooden stocks the spokes are dependent upon each other, and by being tightly wedged together create a mutual support and resistance. The whole are secured by two metal flanges, one at the back and one at the front of the centre of the wheel, which are tightly screwed up, by which means the greatest amount of solidity is obtained for the entire structure of the wheel.

This invention is due to the Messrs. McNeile Brothers, of the Patent Steam Wheel and Axle Company, and it is a significant fact that wheels similarly constructed have for a considerable time been adopted by the Royal Artillery; moreover, they have been extensively used on street cabs, heavy carts, more particularly the latter, and have invariably maintained their character for superiority. For ourselves, we can see that a wheel so constructed must possess peculiar advantages. There is no stock to rot, and the wheel cannot in any sense be spoke-bound, as is frequently the case with wheels of ordinary construction, by the mortise in the stock and the bore in the felloe not ranging in a true line with the spoke. In the growing desire to produce wheels of light construction, great efforts have been made to reduce the size of the centre, and the inventors of these wheels have been very successful in obtaining this object. At the centre their wheels are exceedingly light and ornamental in appearance, and to render them still more uniform they have shortened the arm of their axles, and consequently curtailed the length of the axle-box, so that there is the smallest possible projection at the centre of the wheel. At the same time all the advantages and peculiarities of Collinge's principle are retained. In the ordinary Collinge axle the bearing is not upon the whole length of the arm, and practically speaking Messrs. McNeile have in their axles cut out all that part which is useless in this respect, so that although their axle-arm is considerably shorter, the bearing is the same as in the Collinge axle of ordinary construction.

One of the greatest disadvantages in the manufacture of wheels is the want of uniformity between one another. Scarcely any two wheels are alike. Scarcely any spokes in a wheel radiate alike; some are as much as an inch apart more than others at the felloe; and as the shrinking of the tire varies, some wheels, as a consequence, get more dish than others, the spokes either compressing in the nave mortises, or yielding by elasticity in the direction of their length. To get them at all accurate, it is necessary to employ very skilful workmen, and as skilful workmen are not so numerous as they might be, the cost of wheels is very much increased. Another disadvantage attends them : a workman may put his work badly together, and there is no means of detecting it till the wheel is in actual use. A badly framed wheel will show as well to the eye as a good one, and until it breaks down, no one, whether maker or customer, can detect the inaccuracy. Unless the master watches every wheel while the spokes are driving he can only depend on the good faith of his workmen.

There is no remedy for this evil except substituting machines for men's hands. The machine, if it cuts true once, will cut true always. Every piece of wood in a wheel ought to be shaped by machinery. The felloes should be sawn to their exact size, curve, and length by machine saws ; they should be bored by machine augurs, and rounded by machine shavers. The spokes should be tenoned by machine saws, and shaped by machine lathes. The naves should be cut by a machine lathe, and the mortises in the same cut by a machine chisel. The spokes should not be driven in by the irregular strokes of a mallet, but be forced into their places by the regular pressure of a machine. And when the tire is put on the wheel should be fixed in a frame, in order to preserve an exact size and shape. When all these things are done, we may hope to procure wooden wheels alike in form and quality, and moreover, accurately circular, which very often they are not at present. All the machines should be worked by a steam engine. There is scarcely any article of manufacture for which there is so large a demand, and there is no great variation in their mode of construction. Coachmakers generally seem to cling to the old traditions of their craft with great tenacity ; possibly they think it savours of sacrilege to let progress enter their workshops too rapidly.

The above remarks may be qualified by stating that some of the largest manufacturers have introduced machinery, generally, into the departments in which it is applicable, and

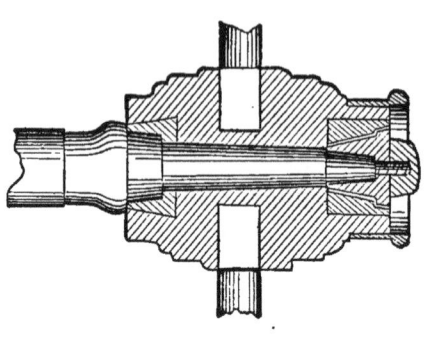

more particularly in the wheelwrights' department. Some years since Messrs. Holmes, of Derby, had mechanical appliances worked by steam power for the following purposes :— Cutting tenons on the spokes, squaring the ends of the felloes, also regulating their length according

Fig. 18.

to the size of wheel required; a narrow upright saw for cutting curved timber ; machine for cutting felloes of the required size and curve ; machine for boring felloes for the

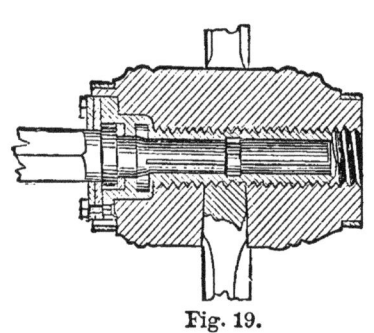

ends of spokes, and many other appliances for lightening hand-labour and insuring greater accuracy in the manufacture. But workshops filled up in this way are not yet the rule, though their number is increasing, and so are the inventions for application of mechanical power to the various processes.

Fig. 19.

It seems rather paradoxical to state that the dished or conical wheel is the strongest. But the fact is, its strength arises from the solid hoop tyre ; with a strake tyre the upright wheel would be the strongest. When running, the great lateral strain on the wheel is from the outside. Consequently, if the wheel be dished in an opposite direction, the thrust will be in the direction of the greatest resistance. The spokes cannot yield, because in yielding they would increase the area of the circle, and this the tyre will not permit. Upon the same principle in carpentry, which constitutes the curved or cambered beam the strongest, the dished wheel is stronger than the straight one.

Here is one very important item which must not be overlooked in the wheelwright department, and that is, the size of the axle-box. The axle-box is a lining of cast iron, on which the axle-arm takes its bearing. Two forms of these are given in Figs. 18 and 19.

Fig. 20 shows an improved form of stock. It will be seen it is to be applied to straight wheels, and requires no further

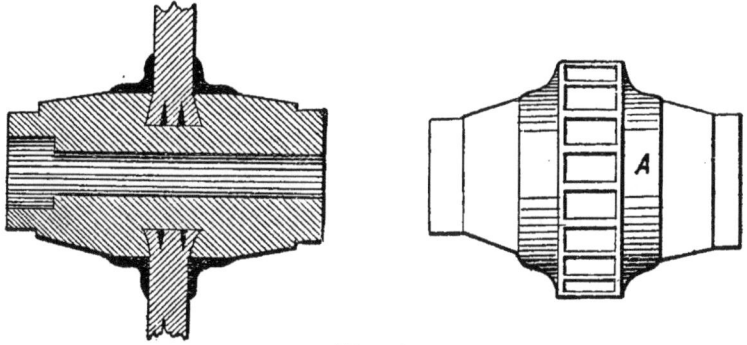

Fig. 20.

description beyond noting that the stock is not weakened so much as in the ordinary way by mortising, an iron band, A, circumscribing the nave and forming a hold for the spokes.

Wheels should be made with a sufficient number of spokes to properly divide the space at the felloes, and afford sufficient support to prevent sinking in between the spokes, and at the same time avoid too many to weaken the stock. The less the number of spokes, the stronger the hub and the weaker the felloe. Judgment should be used in dividing the difference, so as to make each part of the wheel strong in proportion.

CHAPTER VII

AXLES.

An axle, or an axletree, for a locomotive wheel vehicle, is that portion of wood or metal, or both combined, which serves as the axis or centre for the wheels to turn round on.

The name axle-*tree* at once indicates the substance originally employed for it, viz. wood. Axletrees are of two kinds; those which are fixed firmly in the wheels and revolve in gudgeons beneath the wheels, and those in which the wheel moves independently of the axle. The former, as being the rudest, was probably the first kind used. The earliest fixed axletrees were simply pieces of hard timber, with the ends rounded down into a conical form, that form being the easiest to fit to the wheel. Subsequently they were plated with iron to resist wear.

In the earliest iron axles the conical form was still preserved, for the obvious reason of easy adjustment to the wheel. These iron axles were not made in a solid piece, but were merely short ends bedded in and bolted to a wooden centre. Examples of these axles may still be seen in heavy carts and waggons.

The next improvement was to make the axles of a single bar of iron, and this practice has now become common. An axle is technically divided into three parts—the two *arms*, or extremities, on which the wheel revolves, and the *bed*, or that portion which connects the two arms together. The commonest axles, which are manufactured for the sake of greater cheapness, are formed of a square bar simply rolled to shape between mill rollers. This iron is uncertain in its quality, as it is liable to have sand cracks, blisters, and other imperfections, which cause axletrees when made from it to break down under strong concussion. To guard against this, the best axletrees are formed of several flat bars or rods of iron welded together in a mass; this is technically called "faggoting." If you wish to discover whether an axle has been made in this way heat it to a red heat, and if it has been faggoted the grain or lines of the rods of iron running in different directions will be plainly discerned. The size is regulated by the weight it is intended to carry.

For a very heavy coach from 2 to $2\frac{1}{4}$ inches in diameter and 10 to 11 inches long in the arm is a fair size. For light carriages, both four and two-wheeled, $1\frac{1}{2}$ inches in diameter and 8 inches length in the arm is a common size. Occasionally some are made as small as $1\frac{1}{4}$ inches in diameter. It should be remarked that a less size of axle would perform the work required of it if it were stationary, as in mill-work; but for locomotive vehicles it is necessary to provide against the greatest concussion they can meet with in ordinary application.

When iron axles were first used it was customary to drive an iron ring or hoop, 2 or 3 inches broad, into either end of the nave, to prevent too rapid wear. This plan is still used occasionally in heavy carts, but otherwise axles are always fitted with iron boxes, adjusted to the arms with more or less accuracy, according to the price and the material used for lubrication. For the prevention of friction in wooden axles soap or black-lead is the best materials; for common, coarse axles, a thick unctuous grease is the best adapted; but for axles that are accurately made and fitted to the boxes there is no lubricating material equal to oil of the purest kind which can be prepared, i.e. freest from mucilage or gelatine, according as it may be of vegetable or animal production.

The commonest axles now used are of a conical form, with a box of plate iron fitted to them. This box is made by welding the two edges of the iron together in a broad projecting seam, which helps to secure it to the nave. The inside of the box is sunk into hollows for the purpose of holding the lubricating grease. At the upper end of the arm the axle is left square, and against this a large iron washer is usually shrunk on hot. Against this washer the box works. To secure the wheel against coming off a small iron collar is placed on the reduced outer end of the arm, and a linch-pin is driven through the arm beyond it.

An improvement on this kind of axle is when the collar at the upper end or shoulder is made solid by welding, and a screw nut with a linch-pin through it is substituted for the collar and linch-pin. These nuts are commonly made six-sided, with a mortise or slot for the linch-pin through each side, in order to afford greater facility for adjustment. In all other particulars this axle is the same as the last, except that it is occasionally case-hardened to prevent wear and friction.

In travelling, these axles require to be fresh greased every two or three days, and the trouble thus caused is very considerable, besides the risk of omission, in which case the axle is likely to be entirely spoiled.

The commonest kind of oil axle is called the " mail," because the peculiar mode of fastening was first used in the mail coaches. The arm is not conical, but cylindrical, in the improved kind. At the shoulder of this axle a solid disc collar is welded on for the box to work against. Behind this shoulder collar revolves a circular flange-plate of wrought iron, pierced with three holes corresponding with holes in the wheel from front to back, through which long screw-bolts are driven, and their nuts screwed sufficiently tight against the circular flange-plate to allow easy motion. The wheel, when in motion, thus works round the shoulder collar, while the flange-plate secures it against coming off. This is not neat or accurate, but it is simple and secure, and no nut or linch-pin is required to the axle in front, while the front of the nave can be entirely covered in. When screwed up for work, a washer of thick leather is placed between the shoulder collar and the box, and another between the shoulder collar and the circular disc, which extends over the whole surface of the back of the nave. The box of this axle is of cast iron. The front is closed with a plate of metal, between which and the end of the axle-arm a space is left of about 1 inch as a reservoir for oil, which is poured in through a tube passing through the nave of the wheel and closed by a screw pin. At the back of the box there is a circular reservoir for oil, $\frac{3}{4}$ inch in depth and $\frac{1}{2}$ inch wide. When the wheel is in motion the revolving of the box keeps the lubricating material in circulation between the two reservoirs; any portion getting below the arm at the shoulder gradually works its way out and is wasted. The oil in the back reservoir does not waste by leakage so rapidly as that in the front; but when the leather washer becomes saturated with water the oil is liable, by reason of its lightness, to float on the water in or about the washer, and thus get wasted.

This axle requires frequent examination when very much in use; but as it is neat in appearance, and under ordinary circumstances tolerably safe in working, and is not very expensive, it is much used. Both axle-box and axle-arm are case-hardened.

The other kind of axle used by carriage-builders is that known as "Collinge's Patent." The original intention of the inventor was to make it a cylindrical arm, with the box running round it against a coned shoulder, and secured by a coned nut in front; but, as it was found in practice that a leather washer was necessary at the shoulder to prevent jarring, this part of the plan was abandoned.

E

The commonest form of this axle now in use consists of a cylindrical arm with a broad shoulder collar. The box is of cast iron, and the back of it is similar to that of the mail axle before described. The front of it has a rebate cut in the box to receive a small conical collar and the screw of an oil cap. The arm of the axle is turned down in the lathe to two-thirds of the total thickness from the point where the rebate of the box begins. A flat side is filed on this reduced portion, and along it is made to slide a small collar of gun metal, with a conical face in the interior to fit against the coned interior of the rebate in the box. Against this collar, technically called the "collet," a nut of gun metal is screwed, and against that again a second nut of smaller size, with a reversed thread, is tightly fixed. These two nuts, thus screwed in different directions, become as firm as though they were part of the axle itself, and no action of the wheel can loosen them, because the collet, which does not turn, removes all friction from them. But, as a further security, the end of the axle-arm projects beyond the farthest nut, and is drilled to receive a spring linch-pin. Over all a hollow cap of gun metal is screwed into the end of the box. This contains a supply of oil for lubricating purposes.

When the wheel is in motion the oil is pumped upwards from the cap and passes along the arm to the back reservoir, constantly revolving round the cap with the wheel. If the cap be too full of oil—that is, if the summit of the column of oil in the cap be at a horizontal level above the leakage point at the shoulder—it will pump away rapidly, and be wasted till it comes to the level of the leak, where it will be economically used. It is essential to the perfection of an oil action that the oil should not be permanently above the level of the leak, but that small portions should be continually washing up into that position by the action of the wheel in turning.

In order to insure their greater durability and freedom

from friction these axles and their boxes are always case-hardened, *i.e.* their rubbing surfaces are converted into steel to a trifling depth by the process of cementation with animal charcoal for about two hours, when they are plunged into water. The boxes are ground on to the arms with oil and emery, either end being applied alternately, until a true fit between the two is accomplished.

The mode in which oil acts as a lessener of friction is by its being composed of an infinite number of movable globules, over which the fixed surfaces of the arm and box roll without causing that friction and wearing away which would be the result of the two iron surfaces worked together without any lubricant. This saving in the wear and tear of the axle-arm is accomplished by the destruction of the oil. From this we deduce that the greater the mass of oil or grease used the longer will the axle run, and in order to facilitate this as much as possible there should be so much space left between the bearing surfaces of the arm and the box as will allow of a film of oil to be between them.

A highly polished surface is desirable in an axle and box, as the bearing is more perfect and true. A rough surface is a surface of sharp angles, which will pierce through the oil and cause friction by contact.

To guard against the axle running dry, the arm is reduced in thickness at the centre for about an inch to allow a lodgment for the oil, and in the process of working this constitutes a circular pump, which draws up the oil from the front cap and distributes it over the area of the arms. But this, of course, will soon run dry, so that the bèst remedy to prevent the oil being exhausted and the sticking of the axle-arm in the box is careful attention.

A danger arising from careless fitting is the introduction of grit into the box. This grit is composed of small grains of silex, which is very much harder than iron or steel; the consequence is that it cuts and scores the bearing surfaces in all directions, and keys them firmly together, so that it is sometimes necessary to break the box to pieces in order to get it off the arm.

A patent was taken out to remedy these defects by casting three longitudinal triangular grooves in each box. The advantages gained by this are, that if grit gets in it finds its way to the bottom of the grooves and does not interfere with the action of the wheel, and, moreover, the grooves keep up a constant surface of oil in contact with the arm, instead of trusting to the mere capillary attraction. This does not interfere with the bearing surface in any marked degree.

In order that the axle shall be perfect the following considerations are necessary :—

That there be sufficient bearing surface for the arm to rest on.

That the box be of a convenient shape for insertion in· the wheel.

That as large a body of oil as possible be kept in actual contact with the arm by washing up as the wheel revolves.

That the column of oil may be in no case above the horizontal level of the leakage point while the wheel is at rest.

Welding Steel Axles.

Many axles are now made of Bessemer steel. Generally speaking this is neither more nor less than iron, the pores of which are filled up with carbon or charcoal. The higher the steel the more carbon it contains. If steel be heated it loses a portion of this carbon, and the more it is heated the more it approaches its original state, viz. iron.

The welding of steel axles is said to be considerably assisted by the use of iron filings and borax. This is only true in case the steel should be over-heated, and even then only in degree.

Borax by itself is a very useful adjunct to this process, and it should have a small quantity of sal-ammoniac added, to assist its fusion or melting. The furnace or fire, which is to be used for the welding process, should be clean and free from new coal, to prevent sulphur getting on the steel. Of course, all coal has more or less sulphur in it; but iron or steel cannot be successfully welded when there is much sulphur in the fire, so it is well to be as careful in this respect as possible.

Place the ends of the axles in a clean bright fire, heat to a bright red heat, take them out, lap them over each other, and give them a few smart blows with the sledge. Now well cover them with powdered borax, and again put them into the fire and cover them up with coked coal, give a strong even blast, and carefully watch the appearance of the steel as the heat penetrates it, and see that all parts of the weld are equally well heated. When the heat is raised as high as the steel will safely bear (this knowledge can only be gained by experience, so no rule can be given for ascertaining the degree of heat, as it varies with the quality of the steel) take them out. Have two men ready to use the sledges. Place the axles on the anvil, securing them to prevent their slipping, and while one man places his hammer full on the weld, give the extremity of the lap or weld a smart blow or two, and if it adheres then both sledges can be applied until a true and workmanlike weld is formed.

It sometimes happens that when the axles are heated ready for welding and lapped, a light or a heavy blow, instead of uniting the laps, only jars them apart. This is a sure sign that they have been over-heated, and in this case it will be very difficult to form a weld at all. The only way of getting over this difficulty is to heat it to as high a degree as necessary, and put it in a vice and screw it up; the surfaces will adhere in this way when the other means fail.

Another cause of failure is the too free use of borax. If too much is used, it melts and runs about in the fire, unites with the dirt, and generally blocks up the nozzle of the blast,

causing a great deal of trouble to dislodge. If the blast
is not sufficient, then less heat is generated than is necessary,
and it is impossible to form a good weld unless sufficient heat
is applied.

Steel axles do not find great favour with the trade, although
a large quantity of them are used. They are unreliable,
breaking and fracturing without a moment's warning, whereas
an axle of faggoted iron would only twist under the same
circumstances, and could easily be re-forged and set right
again.

Setting Axles.

Setting axles is giving them the bend and slope required,
in order to fall in with the principles of the dished wheel.
It is chiefly applied to the axle-arm, and this is the most important part, setting the beds being mere caprice.

The great object to be obtained is, to give the arm the
right pitch every way, to make the carriage run easy and as
light as possible, even in the absence of a plumb spoke. All
carriages do not look best, when running, with the bottom
spoke plumb or vertical. In some of the heavier coaches or
carriages more slope or "pitch" has to be given to the arm
to carry the wheel away from the body, so as to bring them
to some specified track, in order to suit some particular
customer, so that we must be governed by circumstances.

There is a patent "axle-set," but it is not of much assistance, for half the smiths know nothing about it, and if they
did it would not be generally used, as the advantages derived
from its use are not equal to the trouble of using it. Besides,
the wheels are not always dished exactly alike, and it would
require adjusting to each variety of wheel; and again, the
wheels are not always (though they ought to be) ready; and
when the smith knows the sort of vehicle he is working upon
he can give his axles the required pitch, within half a degree
or so, and the patent axle-set is, unfortunately, not capable
of being adjusted to an idea.

Fig. 21 shows a contrivance for setting the axles when cold, and consists of an iron bar A, 2 feet 1 inch long, and about 2 inches square at the fulcrum B. A hole is punched through the end to allow the screw c to go through; this hole to be oval, to allow the screw to move either way. At the end of this screw is an eye of sufficient size to go on to

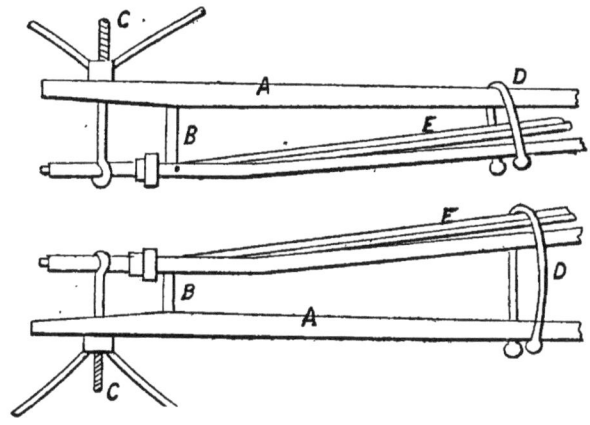

Fig. 21.

the axle-arm. In setting the axle the eye is slipped on to about the centre of the arm; the clevis, D, is placed on the bar A, near the end; the fulcrum, B, is placed at the shoulder, either on top or underneath, according as the axle may be required to set in or out. When the fulcrum is laid on top, a strip of harness leather should be placed on the axle-bed, and on that, an iron E, of the shape of the axle-bed, and on

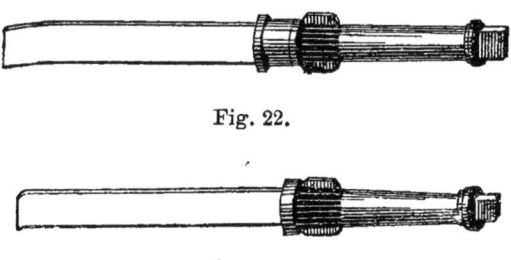

Fig. 22.

Fig. 23.

the end of this the fulcrum is placed; then by turning the screw the axle may be bent or set to any required pitch.

The figure shows the two ways of doing this, one with the bar or lever on top and the other with the lever below.

Figs. 22 and 23 show two improved forms of axles.

Fig. 24 shows another variety of the axle-set. It consists

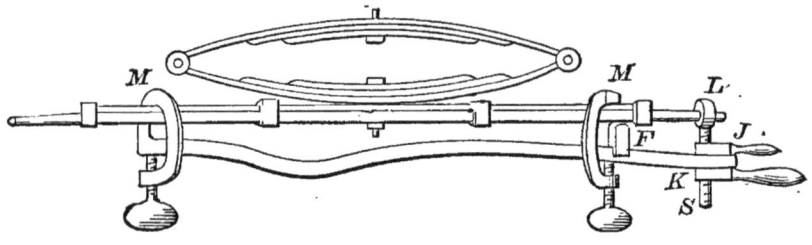

Fig. 24.

of a bar hooked on to the axletree in two places. The bar is fastened by the clamp M, and fulcrum block F. The eye-bolt, L, is hooked over the end of the spindle or arm, and the adjustment of the latter is accomplished by the screw, s, and the nuts J, K.

Weight of Round Iron per Foot.

Diameter. Inch.	lbs.	Diameter. Inch.	lbs.
$\frac{1}{4}$	·163	$2\frac{3}{8}$	14·7
$\frac{3}{8}$	·368	$2\frac{1}{2}$	16·3
$\frac{1}{2}$	·654	$2\frac{5}{8}$	18·0
$\frac{5}{8}$	1·02	$2\frac{3}{4}$	19·7
$\frac{3}{4}$	1·47	$2\frac{7}{8}$	21·6
$\frac{7}{8}$	2·00	3	23·5
1	2·61	$3\frac{1}{8}$	25·5
$1\frac{1}{8}$	3·31	$3\frac{1}{4}$	27·6
$1\frac{1}{4}$	4·09	$3\frac{3}{8}$	29·8
$1\frac{3}{8}$	4·94	$3\frac{1}{2}$	32·0
$1\frac{1}{2}$	5·89	$3\frac{5}{8}$	34·4
$1\frac{5}{8}$	6·91	$3\frac{3}{4}$	36·8
$1\frac{3}{4}$	8·01	4	41·8
$1\frac{7}{8}$	9·20	$4\frac{1}{4}$	47·2
2	10·4	$4\frac{1}{2}$	53·0
$2\frac{1}{8}$	11·8	5	65·4
$2\frac{1}{4}$	13·2		

Weight of Square Iron per Foot.

Side of Square. Inch.	lbs.	Side of Square. Inch.	lbs.
$\frac{1}{4}$	·208	$2\frac{3}{8}$	18·8
$\frac{3}{8}$	·468	$2\frac{1}{2}$	20·8
$\frac{1}{2}$	·833	$2\frac{5}{8}$	22·9
$\frac{5}{8}$	1·30	$2\frac{3}{4}$	25·2
$\frac{3}{4}$	1·87	$2\frac{7}{8}$	27·5
$\frac{7}{8}$	2·55	3	30·0
1	3·33	$3\frac{1}{8}$	32·5
$1\frac{1}{8}$	4·21	$3\frac{1}{4}$	35·2
$1\frac{1}{4}$	5·20	$3\frac{3}{8}$	37·9
$1\frac{3}{8}$	6·30	$3\frac{1}{2}$	40·3
$1\frac{1}{2}$	7·50	$3\frac{5}{8}$	43·8
$1\frac{5}{8}$	8·80	$3\frac{3}{4}$	46·8
$1\frac{3}{4}$	10·2	4	53·3
$1\frac{7}{8}$	11·7	$4\frac{1}{4}$	60·2
2	13·3	$4\frac{1}{2}$	67·5
$2\frac{1}{8}$	15·0	5	83·3
$2\frac{1}{4}$	16·8		

CHAPTER VIII.

SPRINGS.

SPRINGS in locomotive vehicles are the elastic substances interposed between the wheels and the load or passengers in order to intercept the concussion caused by running over an uneven road, or in meeting with any slight obstacle.

A great variety of substances have been used for this purpose, such as leather, strips of hide, catgut, hempen cord, &c.; but these have now been totally superseded by metal springs, so that what is technically understood by the word "spring" is a plate or plates of tempered steel properly shaped to play in any required mode.

It is very probable that the earliest steel springs were composed of only one plate of metal. This was very defective in its action; and unless it was restrained somewhat in the manner of the bow by the string, it was liable to break on being subjected to a sharp concussion.

There is no hard and fast rule by which the spring-maker can be guided so as to proportion the strength and elasticity of his springs to the load they are required to bear; and even were such a rule in existence it would be practically useless, because the qualities of spring steel differ so much that what is known in mathematics as a "constant" could hardly be maintained. The only guide to the maker in this respect is observation of the working of certain springs under given loads, such springs being made of a certain quality of steel, and any peculiar features that appear should be carefully noted down for future reference and application.

Springs are of two kinds, single and double; *i.e.* springs tapering in one direction from end to end, and those which taper in two opposite directions from a common centre, as in the ordinary elliptic spring.

The process of making a spring is conducted in the following manner :—

The longest or back plate being cut to the proper length, is hammered down slightly at the extremities, and then curled round a mandrel the size of the suspension bolt. The side of the plate which is to fit against the others is then hollowed out by hammering; this is called "middling." The next plate is then cut rather shorter than the first; the ends are tapered down so as not to disturb the harmony of the curve. This plate is middled on both sides. A slit is then cut at each end about $\frac{3}{4}$ of an inch in length and $\frac{3}{8}$ inch wide, in which a rivet head slides to connect it with the first plate, so that in whatever direction the force acts these two plates sustain each other. At a little distance from this rivet a stud is formed upon the under surface by a punch, which forces out a protuberance which slides in a slit in the next plate. The next plate goes through precisely the same operations, except that it is 3 or 4 inches shorter at each end, and so on with as many plates as the spring is to consist of. The last plate, like the first, is of course only middled on one side.

The plates of which the spring is to be composed having thus been prepared, have next to undergo the process of "hardening" and "tempering." This is a very important branch of the business, and will bear a detailed description. There is no kind of tempering which requires so much care in manipulation as that of springs. It is necessary that the plates be carefully forged, not over-heated, and not hammered too cold ; one is equally detrimental with the other. To guard against a plate warping in tempering, it is requisite that both sides of the forging shall be equally well wrought

upon with the hammer; if not, the plates will warp and
twist by reason of the compression on one side being greater
than on the other.*

The forge should be perfectly clean, and a good clean
charcoal fire should be used.   Or if coal be used it must be
burned to coke in order to get rid of the sulphur, which would
destroy the " life " of the steel.   Carefully insert the steel
in the fire, and slowly heat it evenly throughout its entire
length; when the colour shows a light red, plunge it into
lukewarm water—cold water chills the outer surface too
rapidly—and let it lie in the water a short time.   Animal oil
is better than water; either whale or lard oil is the best, or
lard can be used with advantage.   The advantage of using
oil is that it does not chill the steel so suddenly, and there is
less liability to crack it.   This process is called " hardening."

Remove the hardened spring-plate from the water or oil
and prepare to temper it.   To do this make a brisk fire with
plenty of live coals; smear the hardened plate with tallow,
and hold it over the coals, but do not urge the draught of the
fire with the bellows while so doing; let the fire heat the
steel very gradually and evenly.   If the plate is a long one,
move it slowly over the fire so as to receive the heat equally.
In a few moments the tallow will melt, then take fire, and
blaze for some time; while the blaze continues incline the
plate, or carefully incline or elevate either extremity, so that
the blaze will circulate from end to end and completely
envelop it.   When the flame has died out, smear again with
tallow and blaze it off as before.   If the spring is to undergo
hard work the plates may be blazed off a third time.   Then
let them cool themselves off upon a corner of the forge;
though they are often cooled by immersion in water, still it
is not so safe as letting them cool by themselves.

After tempering the spring-plates are " set," which consists in any warps or bumps received in the foregoing pro-

* It is the plates that are tempered and hardened, not the *spring*.

cesses being put straight by blows from a hammer. Care should be taken to have the plates slightly warm while doing this to avoid fracturing or breaking the plates.

The plates are now filed on all parts exposed to view, *i.e.* the edges and points of the middle plates, the top and edges of the back plate, and the top and edges of the shortest plate. They are then put together and a rivet put through the spring at the point of greatest thickness, and this holds, with the help of the studs before mentioned, the plates together.

It is evident from the above description of a common mode of making springs, that the operation is not quite so perfect as it might be. The plates, instead of being merely tapered at the ends, ought to be done so from the rivet to the points. And another thing, it would surely make a better job of it if the plates were to bear their whole width one on the other; in the middled plates they only get a bearing on the edges, and the rain and dust will inevitably work into the hollows in the plates, and it will soon form a magazine of rust, and we all know what an affinity exists between iron and oxygen and the result of it; as far as carriage springs are concerned, it very soon destroys their elasticity and renders them useless and dangerous.

To prevent oxidation some makers paint the inner faces of the springs, and this is in a measure successful, but the play of the spring-plates one upon the other is sure to rub off some portions of the paint, and we are just as badly off as ever. A far better plan would be to cleanse the surfaces by means of acid, and then tin them all over, and this would not be very expensive, and certainly protect the plates of the spring longer than anything else.

The spiral springs, used to give elasticity to the seats, &c., are tempered by heating them in a close vessel with bone dust or animal charcoal, and, when thoroughly heated, cooled in a bath of oil. They are tempered by putting them into an iron pan with tallow or oil, and shaking them about over a brisk fire. The tallow will soon blaze, and keeping them on the move will cause them to heat evenly. The steel springs for fire-arms are tempered in this way, and are literally " fried in oil." If a long slender spring is needed with a low temper, it can be made by simply beating the soft forging on a smooth anvil with a smooth-faced hammer.

Setting and Tempering old Springs.

In setting up old springs where they are inclined to settle, first take the longest plate (having separated all the plates) and bring it into shape ; then heat it for about 2 feet in the centre to a cherry red, and cool it off in cold water as quick as possible. This will give the steel such a degree of hardness that it will be liable to break if dropped on the floor. To draw the temper hold it over the blaze, carrying backward and forward through the fire until it is so hot that it will sparkle when the hammer is drawn across it, and then cool off.

Another mode is to harden the steel, as before stated, and draw the temper with oil or tallow—tallow is the best. Take a candle, carry the spring as before through the fire, and occasionally draw the candle over the length hardened, until the tallow will burn off in a blaze, and then cool. Each plate is served in the same way.

Varieties of Springs.

The names given to springs are numerous, but the simple forms are few, the greater part of the varieties being combinations of the simple forms.

The simple forms are the elliptic spring, the straight spring, and the regular curve or C spring (Fig. 25). There are also one or two forms of spring which have become obsolete. Such are the whip spring (Fig. 26), and the reverse curved spring, which was superseded by the last.

The elliptic spring is the one most commonly used at the

present day. Fig. 27, *b*, shows two of these united at the extremities by means of a bolt; this is called a double elliptic spring. The elliptic spring is sometimes used single in what are called under-spring carriages, where the spring

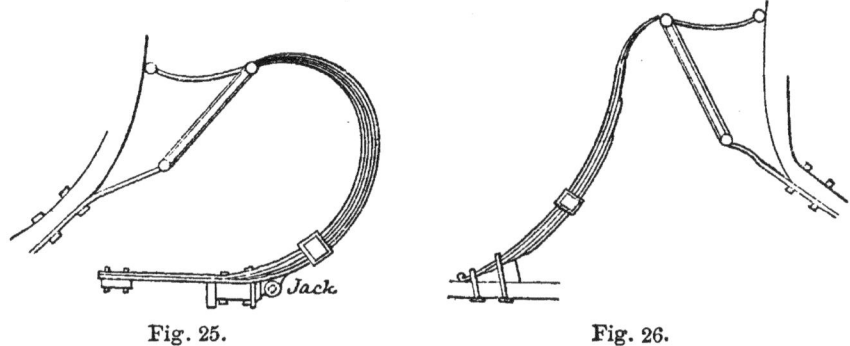

Fig. 25.   Fig. 26.

rests on the axle and is connected with the framework of the body with an imitation spring or dumb iron to complete the ellipse. Its technical name is an "under-spring."

When four pairs of these springs are hinged together so as to form four ellipses they constitute a set, and are used in carriages without perches. Their technical name is "nut-cracker spring."

The straight springs are used in phaetons and tilburies, and are called "single-elbow springs."

The double straight spring is used in omnibuses, carts, &c., where it is fixed across the angle at right angles. It is called a "double-elbow spring."

The regular curved spring is in form generally two-thirds of a circle, one end of which is lengthened out into a tangent, which serves as a base to fix it by in an upright position; the body is suspended from the other extremity by means of leathern braces. Its general figure has caused it to acquire the technical name of C spring. (See Fig. 25.)

The combination known as "telegraph spring" consists of eight straight springs, when used for a four-wheeled carriage, and four springs for a two-wheeled carriage. The

Stanhope is suspended on four of these springs. Two springs are fixed longitudinally on the framework, and two transverse ones are suspended from these by shackles, and

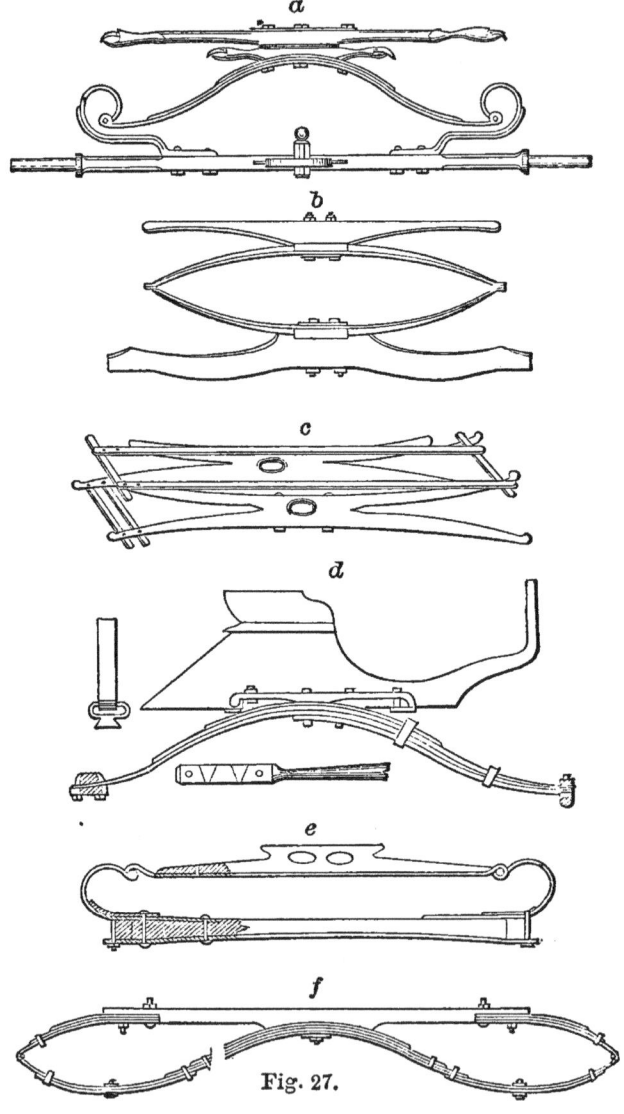

Fig. 27.

on these latter the weight rests. They will bear a great weight, and the body has the advantage of being placed two removes from the concussion.

Fig. 27 shows some varieties of springs.

a Has semi-elliptical springs, hung upon the ends of C springs attached to the axles.

b Has the usual elliptical springs between the bolster and axle.

c Has elastic wooden springs, which connect the axles and support the beds.

d Has some elliptical springs, which also couple the axles A and B.

e Has a bolster hung upon C springs.

f Is a system of curved springs, with three points of connection to the bed and two to the axles.

Weight of Elliptic Springs.

$1\frac{1}{4} \times 3 \times 36$ inch, weight about 28 lbs. per pair.
$1\frac{1}{4} \times 4 \times 36$,,   ,,   34 ,,   ,,
$1\frac{1}{4} \times 4 \times 38$,,   ,,   36 ,,   ,,
$1\frac{1}{2} \times 3 \times 36$,,   ,,   37 ,,   ,,
$1\frac{1}{2} \times 4 \times 36$,,   ,,   41 ,,   ,,
$1\frac{1}{2} \times 4 \times 38$,,   ,,   45 ,,   ,,
$1\frac{1}{2} \times 5 \times 36$,,   ,,   48 ,,   ,,
$1\frac{1}{2} \times 5 \times 38$,,   ,,   51 ,,   ,,
$1\frac{1}{2} \times 5 \times 40$,,   ,,   54 ,,   ,, .
$1\frac{3}{4} \times 4 \times 36$,,   ,,   49 ,,   ,,
$1\frac{3}{4} \times 4 \times 38$,,   ,,   52 ,,   ,,
$1\frac{3}{4} \times 4 \times 40$,,   ,,   55 ,,   ,,
$1\frac{3}{4} \times 5 \times 36$,,   ,,   56 ,,   ,,
$1\frac{3}{4} \times 5 \times 38$,,   ,,   60 ,,   ,,
$1\frac{3}{4} \times 5 \times 40$,,   ,,   64 ,,   ,,
$1\frac{3}{4} \times 6 \times 36$,,   ,,   64 ,,   ,,
$1\frac{3}{4} \times 6 \times 38$,,   ,,   68 ,,   ,,
$1\frac{3}{4} \times 6 \times 40$,,   ,,   73 ,,   ,,
$2 \times 4 \times 36$,,   ,,   58 ,,   ,,
$2 \times 4 \times 38$,,   ,,   62 ,,   ,,
$2 \times 4 \times 40$,,   ,,   65 ,,   ,,
$2 \times 5 \times 36$,,   ,,   63 ,,   ,,
$2 \times 5 \times 38$,,   ,,   67 ,,   ,,
$2 \times 5 \times 40$,,   ,,   72 ,,   ,,
$2 \times 6 \times 36$,,   ,,   75 ,,   ,,
$2 \times 6 \times 38$,,   ,,   78 ,,   ,,
$2 \times 6 \times 40$,,   ,,   85 ,,   ,,

CHAPTER IX.

WHEEL-PLATES AND FORE-CARRIAGES.

THE following is given in the " Coachmaker's Handbook " under the heading of " Short and Easy Turning :"—

" To bring a carriage into a different course from a straight one requires a circular motion, and at half a turn a carriage has established itself in a right angle to its position when at rest.

" A two-wheeled vehicle turns on one wheel, which forms the centre at the place where it touches the ground, and the opposite wheel forms the circle struck from the said centre. The body in this instance follows the circular motion exactly as the axle, and consequently maintains a steady position above the wheels.

" A four-wheeled vehicle remains in a straight line when first the front pair of wheels are turned under, then by the effect of the draught the hind pair of wheels follow in a wider circle. To effect a turning we bring the front axle first in a corresponding direction with the desired turn.

" We make distinction between the moment of turning, or the angular position of the axles previous to the turning itself, and the effected turning of a vehicle round a centre or king bolt, according to the construction of the carriage part. The wheels have to be brought in a position corresponding with the direction of the turning. The body must be fully supported after the turning, and the front or dickey of a carriage must stand in a right angle to fore axletree.

"We have to consider a few points relative to the height of the front wheel, and the elevation of the body above the ground, which averages 30 inches. To give a front wheel its proper height (between 3 feet 4 inches and 3 feet 6 inches) and have it turn a full circle, we sweep the body at the required place, viz. put in the wheel house of a proportioned length, and a depth between 3 and 4½ inches.

"The front carriage part is fastened round the king bolt, turning that part horizontal. This action causes the front wheel to describe a circle, whose diameter is the width of the track; but as the wheel leans over at the top through the dish, we have a larger circle in the middle and top of the wheel. We, therefore, first find a top circle, having a diameter equal to the width between the highest point of the wheel, and a side circle following the termination of the cross diameter of the wheel, having as a centre the king bolt."

In Fig. 28, which is drawn to quarter-inch scale, the horizontal line A is the axletree, B is the wheel at rest, C is the wheel on full lock, D is the back of the arch, E shows the circle that the wheel will describe on the ground as it moves backward, and F is the circle the back of the wheel will describe in the air at the same movement. It will be seen by this that when the wheel is on

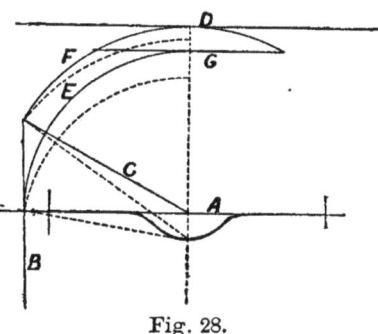

Fig. 28.

half lock the back part of it will come in contact with the arch, and that when on full lock it will have travelled right away from it. It follows, therefore, that if we want to find out the right position for the perch bolt to occupy, we must not measure the circle the wheel will describe on the ground, but the one described in the air. We must, therefore, measure along the line F, and carry that measurement along to D.

The length of the line A to D is exactly 3 feet. Now the

position of the perch bolt, or centre point on which the
wheels lock round, need not be, and in fact very rarely is, in
the same vertical line as the axletree. By compassing the
beds or timbers on which the fore part of the body rests and
through which the perch bolt passes, the centre of the circle
described by the lock may be carried forward. Thus, if the
beds be compassed 4 inches forwards from the straight line A
along the axle, the centre will be carried forward 4 inches,
the result of which will be, that when on the half lock the
back part of the wheel will be carried away from the arch
2 inches, and that when on the full lock the wheel will
stand 4 inches from the position it would occupy if the bed
were straight. The dotted lines below F and E in the figure
show the result of this difference in the shape of the bed ;
and it will be seen that to get a 2-inch clearance of the arch
from the back of the wheel without carrying the wheel itself
farther forward than 3 feet from the back of the arch, we
must compass the bed 4 inches, the compass mark being to
the centre of the substance of the bed.

In the fore-carriages for one-horse vehicles, and two-horse
vehicles as well, the shafts are carried by " open futchells "
(F, Fig. 30) ; and in the fore-carriages of two-horse vehicles
the pole is carried by " close futchells " (F, Fig. 31).

A reference to the figures will make the following remarks
more clear. The central circle is the wheel-plate, or, as the
Americans term it, the fifth wheel. This is flat at the bottom
and round on the top, and being fitted to the under part of
the top carriage takes its bearing on the bottom carriage,
and by its extended circular formation gives steadiness to
the body when the carriage is running in a straight line, or
when the fore-carriage is on full lock. These bearings are at
the back ends and fore part of the futchells, and at those
points of the bottom bed which are covered by the wheel-
plate. The fore and hind bearings are of ash timber, and
are necessarily circular in form.

It will be seen how imperative is the necessity of the wheel-plate being a perfectly true circle and of its taking a perfectly flat bearing; the forging and finishing such a piece of work requires, therefore, peculiar care and skill. In the figure showing the open futchells, the stays which run from the back end to the front of the futchells are the wheel-irons, the back-stays, and the bed-clips in one. They clip over the ends of the bottom bed, and being at these points made flat, they are cranked downwards to take their bearing on the spring-block, and here they are fixed to the springs either by means of bolts passing through them or by clips and couplings. The best plan is by the latter, as when they are bolted on holes have to be drilled through the springs, which renders them weak at these points; the same may be said of the manner of fixing the bottom half of the spring on to the axle.

The wheel-iron, bed-clip, and back-stay being in one, a good opportunity is afforded to the smith to display his skill and taste, as it is desirable that this piece of workmanship should be well forged and fitted, and at the same time a certain grace of outline must be given to it, otherwise the appearance of the vehicle will be spoiled. When it is finished it should fall into its position and take all its bearings accurately without force being used, for if in bolting on any strain should be put upon it, in order to get it into its place, it will be liable to snap on meeting with an obstacle on a rough road.

The English coachsmith ought to possess a better knowledge of metallurgy than he does. All smiths get a certain rule-of-thumb knowledge, but what they should possess is a thorough scientific knowledge of the properties of the metals they are dealing with. We cannot enlarge here upon the subject, but the artisan will do well to study a work like "Metallurgy of Iron," in Weale's Series; and if it only teaches him to tell accurately the good qualities of metal from the bad, he will have a greater knowledge than a large number of coachsmiths have. For general purposes it is as well that he should know that perfectly pure iron is so soft and tough, and at the same time so malleable, that it can be rolled into sheets $\frac{1}{300}$ part of an inch in thickness; and that when wrought iron can be twisted, cold, into almost any shape without breaking, he may rest assured that it is as near pure iron as any one could wish to have for the proper execution of smith's work.

Fig. 29 illustrates a light fore-carriage, with drop pole and shafts, suitable for light phaetons, coupés, and Victorias.

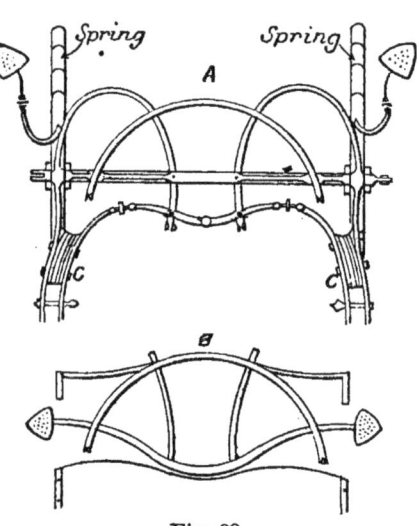

Fig. 29.

The portion marked A represents the lower part. The new mode of constructing this carriage is the doing away with the bent futchells and using puncheons in their places. The inside front-stay is forged in one piece; in the centre is formed the socket to receive the king or perch bolt. The stay rests on the top of the two puncheons. There is a T plate formed solid with these stays running back to the bed, and at c forms the inner part of the socket for receiving the shafts. The back-stay passes around under the puncheons, crossing the bed to the front and bolted where the front-stay crosses the puncheons, the other end extending to the front, forming the outside of the socket for receiving the shafts. o o represents a piece of hickory bolted between the two stays.

The following dimensions applicable to the figure may be useful:—Springs $1\frac{1}{2}$ inches thick, consisting of four plates 37 inches long, $11\frac{1}{2}$ inches opening, which may be varied to suit the body of the carriage they are intended for. Lower bed $1\frac{1}{4}$ by $1\frac{1}{3}$ inches; $\frac{3}{8}$-inch plate on the bottom. Wheel or stay-iron ½ inch round, increasing the size to the puncheons. The box clips over the bottom bed with clip bars, which are worked solid. The clips are put on from underneath the springs, and are se-cured by nuts on the top. The size of the half-wheel iron is 1 inch by ½ inch.

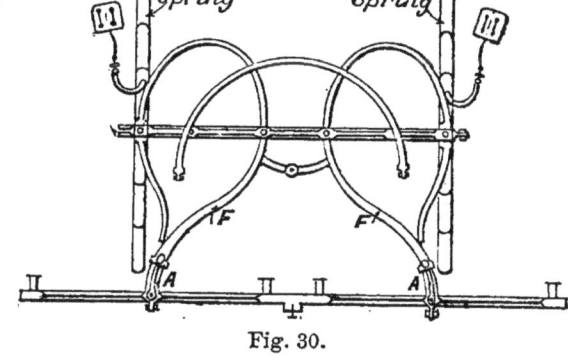

Fig. 30.

The portion of the figure marked B gives the upper por-tion of the fore-carriage.

Fig. 30 shows a fore-carriage with open futchells. A A is where the stiff bar detaches to receive the shaft of the drop poles, the futchells extending to A A. The wheel-iron or stay on the outside of the futchell extends ahead 5½ inches, and also the plate on the inside of the futchells the same distance. These irons require to be a good thickness and tapering to the end. The

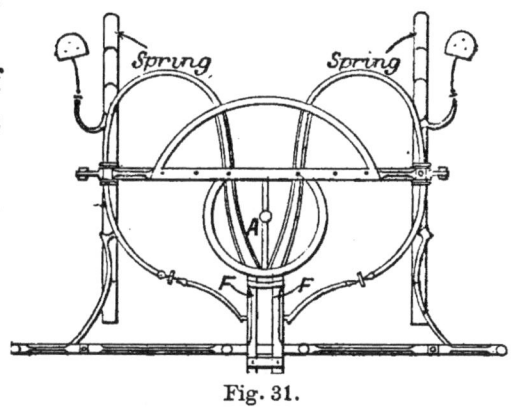

Fig. 31.

blocks are fitted on to this space and scrolled at the end. The dimensions given for Fig. 29 will apply to this, and it is used for the same light vehicles, only that a pair of horses are used instead of one.

Fig. 31 shows an arrangement adapted for hard service. With this kind of wheel-plate we get a good bearing when it is turned under the body. These are made solid, with a plate on the two beds. A is the perch bolt.

CHAPTER X.

IRON AND METAL-WORK GENERALLY.—LAMPS.—PRINCIPLES OF COMBUSTION.

In addition to the foregoing, a large quantity of expensive ironwork is used in the construction of carriages; the principal cause of which expense is not the cost of the material, but the highly skilled labour which is necessary in preparing it.

In carriages suspended from C springs, the front and hind wheels are connected by a central longitudinal timber, called a perch. This has to be plated with iron in order to prevent its breaking when running over bad ground; and without this precaution there is great danger of the perch giving way, as owing to its curved form, which follows the contour of the body, it is necessary to cut across the grain, and thus weaken the timber.

These wooden perches have, to some extent, been superseded by wrought-iron perches, as previously mentioned, and these have been found to answer admirably.

In C spring carriages there are the loops, which serve to suspend the body; and these require very good workmanship, for they are curved in many opposite directions, are tapered and irregularly formed every way, yet requiring to have bearing bolts accurately adjusted, and sundry contrivances for affixing ironwork to them, and all this without a single square side for the mechanic to work from. They are samples of great mechanical skill and dexterity of hand.

and jacks.

Stays are iron brackets of various forms, bolted by their extremities to such parts as they are intended to sustain or strengthen, but they do not take a bearing on any part.

Plates are irons which take a bearing throughout their length and breadth on the part they are intended to strengthen, and to which they are fastened by bolts, screws, or rivets.

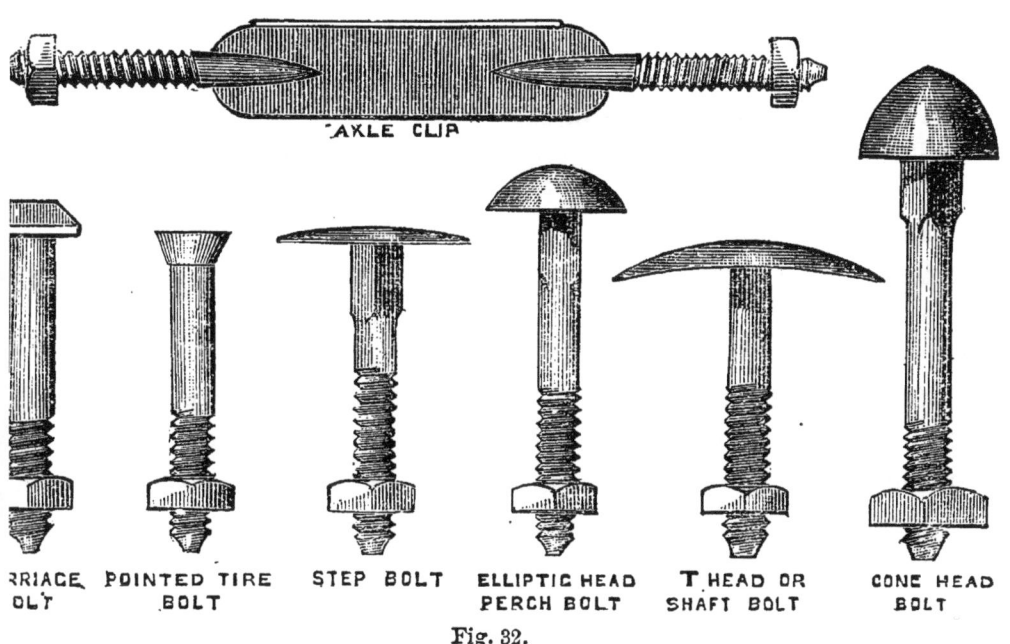

AXLE CLIP

| RRIAGE | POINTED TIRE | STEP BOLT | ELLIPTIC HEAD | T HEAD OR | CONE HEAD |
| OLT | BOLT | | PERCH BOLT | SHAFT BOLT | BOLT |

Fig. 32.

Hoops are flat straps of iron riveted or welded together, for the purpose of securing timbers together side by side.

Clips are a kind of open hoops, the ends of which have a thread run upon them in order to take nuts. The purposes for which they are used is to screw springs and axles in their places without having to weaken them by drilling holes through them.

Bolts are cylindrical pieces of iron of various sizes (Fig. 32),

F

one end of which is flattened out to form a head, and the other is formed into a screw to receive a nut. The use to which they are applied is to secure the ironwork and heavy framework.

Steps may be single, double, or treble. In the two latter cases they are made to fold up, and are called folding steps, and may be made to fold up outside or into the body ; this latter is the best way, and if they be well managed they do not incommode the sitters inside.

Treads are small single steps a few inches square, fixed for the most part on a single iron stem.

Joints are jointed iron stays, made in the form of the letter S, and serve to keep the leather heads or hoods of open carriages, such as landaus, stretched firmly out when required.

Shackles are iron staples, which serve to receive the leather suspension braces of C spring carriages on the springs ; they are also used for coupling springs together.

Jacks are small windlasses, which serve to receive the ends of the leather suspension braces after passing round the backs of the springs. By means of a wrench or winch handle the jacks may be wound up or let down so as to lengthen or shorten the brace.

Then we have the hinges, which are now concealed in the door pillar, effecting a great improvement in the appearance of the vehicle, though it necessitates a somewhat stouter pillar than would otherwise be necessary.

To preserve the ironwork and steelwork of carriages from rusting, it is either painted or plated with some metal on which the oxygen of the air does not act. When it is wished to make it ornamental, carriage ironwork is plated, in which case it is first covered with a coat of tin laid on by means of a soldering iron, with rosin and a small portion of sal-ammoniac in order to promote union between the two metals. The tin being smoothed, a small portion of silver or brass, rolled exceedingly thin, is laid on, and by means of the soldering iron is made to adhere to the tin; more of the plating metal is then added to join the first by the edges, till the whole surface is covered. It is then burnished and polished by means of the proper tools. All articles of iron requiring to be covered with silver are treated in the same way; small articles of ornament in brass, which do not require strength, are cut in solid metal, as it is cheaper by the saving in labour; but for heavy articles, the weight of the metal would considerably enhance the price, supposing that strength were not required. Wheel nave hoops, axletree caps, loops, brace buckles, check rings, and door handles are generally plated.

The beading, which is used to cover the joints, is of three kinds, brass, copper, and plated copper. It is formed by strips of metal being drawn into a circular or angular form by means of a die, the hollow space being filled with solder, into which small pins of pointed wire are fixed to attach it by. The brass beading is polished; the copper is painted, for which purpose the surface is roughened. As the quantity of beading used is often very considerable, the labour of silvering by means of a soldering iron would be too great, and therefore the plated or silvered beading is prepared from metal silvered in the sheet. The process is very simple:—A bar of copper being reduced to the proper thickness, a bar of silver is then united to it by means of heat. They are then passed through the rolls together, and occasionally annealed in the fire until the requisite thickness be obtained, the two metals spreading equally. This kind of metal is much used in the manufacture of carriage lamps.

Several kinds of lamps are used in carriages, both as regards principle of construction and form and ornament. In the simplest kind the light is furnished by the combustion of wax candles, which are contained in tin tubes, through a hole in the upper part of which the wick passes, the candle being pressed upwards as fast as it consumes by a spiral spring. In dress carriages, where the lamps are somewhat ornamental, wax candles are invariably used on account of their superior cleanliness, though the light is inferior to that of oil.

Oil is often used on account of its superior illuminating power. The lamp then simply consists of a tin reservoir for holding the oil, and a round wick of the most ordinary kind, though sometimes flat for the sake of spreading the flame. Reflectors of many kinds are used in every variety of carriage lamps, formed of silvered metal highly burnished.

Attempts have been made to bring the argand lamp, with a current of air through the wick, into use, but sufficient success has not attended these efforts to make them general, owing to their liability to be suddenly extinguished by violent draughts of wind. This may, however, be accomplished when the scientific principles of combustion and the regulation of the draught shall be better understood.

By a common lamp is understood one that feeds the wick with oil by capillary attraction, the column of oil being below the level of the flame. An argand lamp, on the contrary, has a column of oil considerably above the level of the flame and constantly pressing upwards to it like a fountain. The motion of a carriage has a tendency to make oil at times flow too rapidly and extinguish the flame, and sometimes to cause too sudden a rush of air up the central tube, which blows away the flame from the wick; and when these difficulties are overcome by ingenuity they become such complicated pieces of work that it is beyond the ordinary " gumption " of the servants to trim them properly, and if this be not done the object of the improvements is defeated.

The principle of constructing an argand lamp, so that it may regulate its own air draught, is set forth by Lord Cochrane in one of his patents. It is to divide the lamp into three chambers—one in the centre, which contains the reflectors and light, surrounded by the chimney glass, and is pierced with holes at the sides to permit the egress of the heated air and the ingress of the fresh air. From this chamber a tube or tubes communicate with the lower chamber, into which the air tube of the burner descends, and thus furnishes a regulated supply of air. It is evident by this process the air rushing in must be regulated by the air rushing out, and *vice versâ.*

The flame of the lamp is not produced by oil or tallow alone, it requires the oxygen of the air to mix with it in order to sustain combustion. This may be proved by putting a glass bowl over a candle, when, as soon as the oxygen is consumed, the light will be seen to go out, the bowl having the remaining constituent of air left in it, viz. nitrogen, which will neither support nor assist combustion. Herein consists the advantage of the argand lamp in furnishing atmospheric air in the centre of the flame. The flame arising from a thick wick is hollow, *i.e.* it is a film of light, like a bladder, and not continuous, the inner portion of the flame being filled with gas.

It is well known that if the flames of two candles be brought in contact they will produce a greater intensity of light than if burned separately. Upon this principle what are technically termed "cobblers' candles" are made. For the same reason lamps are sometimes made to burn two or three wicks, placed just so far apart that the flames may come in contact. This is an approximation to the argand principle, by admitting air between them. But there is one difficulty attending them, viz. the regulating of all the wicks to an equal height, which would be considerable unless the lamp were so contrived that all could be regulated by one movement. If this difficulty can be overcome a very excellent lamp might be made by placing four wicks in a square.

CHAPTER XI.

PAINTING.

PRINCIPLES OF COLOURING IN PAINTING.

COLOURS are distinguished by artists as *pure, broken, reduced, grey, dull,* &c.

The pure colours consist of those which are called simple, or *primary;* these are red, yellow, and blue; and those which are formed from their mixture in pairs (binary compounds) are termed *secondaries:* such are orange, violet, green, &c.

Broken colours are formed by the mixture of black with the pure colours, from the highest to the deepest tone.

A *normal* colour is a colour in its integrity, unmixed with black, white, or any other colour.

The mixture of equal parts of red and yellow produces orange; mix equal portions of yellow and blue and we have green; equal parts of red and blue produce violet. These are called *secondary colours.*

Three parts of red mixed with 1 part of blue produce violet-red; 3 parts of red and 1 of yellow produce a red-orange; 1 part of red to 3 of yellow produces orange-yellow; 3 parts of yellow to 1 of blue produce a light yellow-green; 1 part of yellow to 3 of blue produces a blue-green; 1 part of red to 3 of blue produces a light violet colour. All these are called *secondary hues.*

Normal grey is black mixed with white in various proportions, producing numerous tones of pure grey.

Lamp and gas-lights throw out yellowish-coloured rays,

causing a great many light colours to appear different in tone from what they really are. Certain shades of green and blue are not easily distinguished by gaslight. A blue fabric will appear to be green, or of a greenish tone, caused by the yellow rays falling on it. Green being formed by the mixture of blue and yellow, whatever contributes yellow to blue, as in the case cited, or by mixture of pigments, the hue will be green.

When coloured rays fall on a coloured surface, which is lighted by diffused daylight, the coloured surface is changed, the effect being the same as that produced by adding to it a pigment of the same colour as the coloured light. When red rays fall upon a black stuff they make it appear of a purple black; on white stuff they make it appear red; yellow stuff they make appear orange; and light blue stuff they make appear violet.

Complementary Colours.

" The colour required with another colour to form white light is called the complementary of that colour. Thus green is the complementary of red, and *vice versâ;* blue is the complementary of orange, and *vice versâ;* yellow is the complementary of violet, and *vice versâ*, because blue and orange, red and green, yellow and violet, each make up the full complement of rays necessary to form white light."

These remarks are deduced from experiments with a prism of glass, giving the spectrum or analysis of the coloured rays forming white light. When a ray of sunlight is passed through a triangular prism of flint glass, and the image received on white paper, it will be noticed that the spectrum (as the image is termed) consists of several colours—seven in all. Red, yellow, and blue are the most prominent. The red rays are modified by the pure yellow, and we have orange; the yellow rays becoming mixed with the blue become green; the blue and red give violet.

Look intently for a few moments at a bright red object, then suddenly transfer the gaze to a sheet of white paper; the paper will appear of a greenish tint. Reverse the process; look intently at green, then on white paper, and red will be the tone of the paper. Blue will excite the eye to see orange, and orange will excite the eye to see blue. This is called successive contrast. In placing colours near each other it is of the greatest importance that the painter should bear in mind the foregoing laws. From these laws the coach-painter may derive some useful hints. In ornamenting and striping bear in mind that colours that are complementary purify each other.

The effect of placing white near a coloured body is to heighten that colour. Black placed near a colour tends to power the tone of it. Grey increases the brilliancy and purity of the primary colours, and forms harmonies with red, orange, yellow, and light green.

Chiaro-oscuro and Flat Tints.

"There are two systems of painting; one in chiaro-oscuro and the other in flat tints. The first consists in representing as accurately as possible upon the flat surface of canvas, wood, stone, metal, &c., an object in relief in such manner that the image makes an impression on the eye of the spectator, similar to that produced by the object itself. Therefore every part of the image which receives in the model direct light, and which reflects it to the eye of a spectator viewing the object from the same point in which the painter himself viewed it, must be painted with white and bright colours; while the other parts of the image which do not reflect to the spectator as much light as the first must appear in colours more or less dimmed with black, or what is the same thing, by shade.

"Painting in flat tints is a method of imitating coloured objects, much simpler by its simplicity of execution than the preceding, which consists in tracing the outline of the different parts of the model, and in colouring them uniformly with their peculiar colours."

Paint Shop.

The paint shop should be a roomy apartment, well lighted and ventilated. If possible, bodies and carriage parts should have separate shops to be painted in ; the rough work on bodies, too, ought to have a separate room for its execution.

There should be a good assortment of brushes suitable for every variety of work, plenty of paint pots, at least two paint mills, marble slab for mixing colours on, and a stone to be used exclusively for making putty on ; water buckets, sponges, chamois, palette knives, and putty knives. Light trestles set on casters for light bodies, and heavy trestles, with two wheels and a pole, for heavy bodies.

Screens, covered with heavy paper or enamelled cloth, to protect varnished work from floating particles of dust, and also from the unsightly marks left by flies, will be found very useful.

Also the necessary colours ; white-lead, whiting, ground and lump pumice-stone, &c.

The brushes used are of various sizes and have various names. The largest are used for covering large surfaces with paint. A smaller kind are called *tools*, or *sash tools*, the name *tool* being applied more especially to the smaller varieties. Those brushes used for striping and ornamenting, being very small, are called *pencils*. All these are made round and oval, and filled with various kinds of bristles.

There are also *flat* bristle brushes of various sizes, which are useful for body painting. A small variety of these is also used for painting the carriage parts.

For varnishing, the black sable and badger brushes (both flat) are handy, though in applying rubbing varnish, the hair of these is sometimes too soft to lay it on without its being thinned with turpentine. But they are capable of producing very finely finished work.

For painting carriage parts, the medium-sized brushes should be used. For lead and rough coatings on bodies a larger brush is required than for carriage parts. Body brushes should be kept separate from those used for carriage parts, as the latter wear the brush hollow in the centre, which unfits them for laying a level coat on the body.

The *pencils* are made of sable, camel, and cow hair. Sable-hair is of two colours, red and black, either of which is superior to any hair now used. The red sable-hair is rather finer than the black, and is rather better adapted for ornamenting, while for striping and lining the black sable is very suitable. Camel-hair pencils work very well for broad lines. Those made from cow-hair are not much used.

Ornamenting pencils are either made in quills, or tin-bound with handles. They are of various sizes, suited for the most delicate touchings or broad handlings. They should be kept clean, as the smallest quantity of dry paint in them prevents them working well. When not in use, they should be greased and put away in such manner that the points may not be bent.

Lettering pencils, of sable and camel hair, are commonly used; they should be from $\frac{1}{2}$ inch to 1 inch long, the shorter ones being used for filling in after the outlines are traced.

There are three or four kinds of paint mills manufactured. They vary in size and price, and are suited to the wants of large or small factories. Where several painters are employed there should be two paint mills, one kept exclusively for grinding colours, and the other for lead, filling, pumice-stone, &c. By this means the colours are not so likely to be soiled by mixture with lead colours or other rough heavy paints, which clog the mill up so rapidly.

Paint mills have not penetrated into every workshop yet. In some is still used the old-fashioned slab and muller; this is a more tedious operation, and is not so successfully accomplished as by the mill.

Colours.

The colours generally used in the carriage paint shop are the following:—White-lead, whiting. Yellow ochre and red-lead, used for rough work. *Ground colours*, which the painter uses in combination with other pigments, as chrome yellow, Indian red, raw umber, Prussian blue, &c. *Panel colours*, as carmine, lake of various hues, ultramarine blue, verdigris, &c.

White-lead is very largely used, not only as a foundation, but enters into the composition of various colours, as stone, drab, straw, &c. In the mixture of rough stuff or filling, white-lead gives elasticity and life to the ochre, and when properly used forms the tenacious part of the under-coatings. But oil white-lead should not be used where there is not sufficient time for it to thoroughly dry.

After a good foundation has been secured, and smooth coatings of lead are desired, which will sand-paper smoothly and leave a pleasant surface to colour over, the dry or tub white-lead should be used. Whiting and white-lead make a good putty, though it is not so much used as it ought to be.

CHROME YELLOW is seldom used clear, except for line-striping. There are different shades of it as well as qualities, the best being the cheapest in the end, as it has more body. Lemon and orange chrome are all the carriage painter requires; with these he can mix up any hues needed by the addition of reds.

INDIAN RED is a strong colour and of great service to the painter, especially in forming the groundwork for transparent colours, such as lakes of a reddish or purple cast, and carmine. Mixed with lampblack, it forms the most durable under coatings that can be obtained where a brown is needed.

Raw Umber is largely used. With blue and yellow it forms a pleasant range of quiet greens.

Combined with white and yellow it gives drab tints or stone colour, which may be toned down by adding black, or lightened up by vermilion or lake. In mixing a light striping colour which may have too much of a raw yellow tone, if a little umber be added the defect is corrected.

Carmine is a very brilliant colour, surpassing vermilion in richness of tone, and yet similar to it for height of colour. It is often adulterated with vermilion, which of course injures its purity. *Pure* carmine will dissolve in ammonia water without leaving any sediment.

Mixed with drabs, delicate greens, asphaltum, &c., carmine imparts warmth without injuring the colours.

There are several tints and qualities of Lake. Those commonly used are English purple, Munich and Florence lakes. English purple lake will bear some raw oil in mixing it; the others are best without it.

Ultramarine Blue, when pure, is a very durable colour. It is prepared from the mineral *Lapis lazuli*. Mixed with the lakes, it tones them down without seriously injuring the purity of the colour. For clear ultramarine a dark lead-coloured ground will answer, or make a ground colour of Prussian blue and white to nearly match the tint of the ultramarine.

This is rather a difficult colour to handle, but the secret of laying it on successfully is to have sufficient varnish or boiled oil in the colour to prevent it " flying off " or drying too dead.

Verdigris is an acetate of copper, of bluish colour, and requires a groundwork. It is not used to any great extent as a panel colour.

The following is a list of compound colours and their application :—

Pure Grey.—White and black in various proportions.

Coloured Greys.—Red and green. Blue and orange.

Straw Colour.—White, chrome yellow, and raw umber.

Light Buff.—White and yellow ochre.

Deep Buff.—White, yellow ochre, and red.

Salmon Colour.—White, yellow, and vermilion.

Flesh Colour.—White, Naples yellow, and vermilion.

Orange.—Equal parts of red and yellow.

Pearl Colour.—White, black, and vermilion.

Lead Colour.—White and blue, with a little black.

Stone Colour.—Black, amber, and yellow.

Canary Colour.—White and chrome yellow.

Tan Colour.—Burnt sienna, yellow, and raw umber.

Pea Green.—White and chrome green.

Sea Green.—Prussian blue and yellow.

Citron.—Green and orange.

Chocolate.—Black and Spanish brown.

Olive.—Umber, yellow, and black.

Lilac.—White, carmine, and ultramarine blue.

Purple.—Olive, red, and carmine.

Violet.—Blue and red.

Wine Colour.—Purple, lake, and ultramarine blue.

Dark Brown.—Vandyke brown, burnt sienna, and lake.

Green.—Blue and yellow in different proportions according to the tone required.

Marone.—Crimson, lake, and burnt umber.

The above list will enable the painter to mix about all the colours required in coach-painting. A great many shades may be made of each of those given by varying the proportions of the component colours.

A good quality of raw linseed oil should be used, as it works well and dries dead when not used in excess, and it is free from the gumminess found in boiled oil. Raw oil simmered over a gentle fire for two or three hours has its drying qualities improved, more especially if a little sugar of lead be added to it.

Japanners' gold size is made as follows :—Asphaltum, litharge, or red-lead, each 1 oz.; stir them with a pint of linseed oil, and simmer the mixture over a gentle fire till it is dissolved and the scum ceases to rise, and the fluid thickens on cooling.

The quality of the varnish used is very important. Rubbing varnishes are required to dry firmly in from two to five days, consequently they have not much oil in their composition. A good wearing rubbing varnish should not be rubbed until the fourth or fifth day after being laid on ; when rubbed it should not sweat (become glossy) soon after, even in hot weather. Slow-drying rubbing varnish, when allowed to stand a day or so after having been rubbed down, will sweat out in hot weather, and should again be run over with the " rub rag " and fine pumice before another coat is applied.

Rubbing varnish that sweats at all times, soon after being rubbed, is liable to crack and should not be used.

By the use of hard drying varnish the painter is enabled to level his work down, and prepare for the last coat or finishing varnish. This last coating must be of an opposite nature to that on which it is laid if great brilliancy is sought after ; and, as its surface must ever be opposed to the action of heat and cold, sunshine and shower, it must possess an elasticity or oily nature that will resist these changes for a great length of time.

Painting the Coach.

The body generally receives a coat of priming on the bottom, top, and inside in the wood shop. This is called " slushing," so that when the body arrives in the proof shop these parts are one coat in advance of the other portions.

The top and the panels require a considerable amount of attention. If the top is constructed of green timber it will cause the covering to rise up in ridges or blisters, and when the canvas is put on, if it is not well stretched when it is nailed on, the air gets under it and causes a deal of trouble. The inside of the top should have a good heavy coating of slush or oil lead to preserve the wood from dampness, and the outside of the top should be properly primed with clean smooth lead colour. When this is dry the nail holes should be puttied in, and sunken places brought forward with firm drying putty, which will bear blocking down with sand-paper, leaving the top as level as possible. When this is dusted off clean, apply a heavy coat of lead in oil, with sufficient varnish in it to hold the lead together.

The inside of the body should be well coated, as it is a great protection to the panels. The priming coat should be composed of the best pure keg lead and oil, with only a small quantity of drier, and allowed at least a week in which to dry. This coat should be well worked in to the nail holes and the grain of the wood. A well-worn springy brush is the best. When this coating is dry sand-paper it carefully, and apply a second coat of lead colour, using less oil. The third day after this, putty the nail holes half full; two after this, apply the third coat, mixing it so that it will dry firmly, no oil being used except that which is present in the lead. When dry, finish puttying the nail holes; also putty up any of the grain that may appear too open, or else rub into the grain some lead mixed up very heavy.

The body, after having received three coats of lead, and been puttied up, may now stand for two or three days. When it is again taken in hand, sand-paper off any putty that may be above the level of the surface; dust off, and brush on a level coat of lead, which must dry hard and firm. Every coat of lead should be laid on as level as possible, and made to fill up the grain of the wood as much as possible. These coats are called "rough-stuff." The body may now stand for three or four days, when it will be ready for the filling up.

There are two very important things to be studied in coach-painting. First, to form a surface hard enough to hold out the varnish and disguise the grain of the wood; and second, to have the first and intermediate coats of paint sufficiently elastic to adhere and yield to the natural action of the wood without cracking or flaking off. In effecting one of these results we are apt to affect the other; and nothing but the utmost care, on the part both of the manufacturer of the essential ingredients and of the person who prepares them for use, can insure durability.

The leather-covered portions are usually primed with two coats of black Japan, reduced with a little turpentine.

A good stopping material for nail holes, &c., is made of dry lead and Japan gold size. It is called "hard stopper."

The rough coatings should dry firmly, possessing only sufficient elasticity to bind them to the surface. The first coat will bear a trifle more oil than the remaining ones, and should stand about four days before the others are put on, which can be done every other day.

Five coats of filling up are next added, composed as follows :—

> 2 parts filling-up stuff.
> 1 part tub lead.
> 2 parts turpentine.
> 1 part Japan gold size.
> $\frac{1}{2}$,, bottoms of wearing varnish.

The first coat should cover every portion of the lead surface, be well brushed in, but not allowed to lie heavy at the corners. The remaining coats may be applied reasonably heavy, but kept from lapping over the edges or rounding the sharp corners, and thus destroying the clean sharp lines of the body-maker.

Any defects noticed while filling in should be puttied or stopped, ever bearing in mind that the perfection in finish aimed at is only secured by care at every step taken.

The leather-covered parts generally have three additional coats of filling in.

The time allowed for each coat to dry may be extended as far as convenient, but there is nothing to be gained by allowing weeks to intervene between the coatings. When a coat is hard it is ready for another; and it is far better to have the body filled and set aside than to divide the time between the coatings, and probably be compelled to rub out the body before the last coat is firm. Of course, the time occupied by the coats of filling in to dry varies according to its composition. If much oil be used, it will take a longer time for each coat to dry; but the above composition may be applied one coat every other day.

The first coat may be applied rather thinner than the others, and is improved by being mixed with a little more white-lead. It should also be made more elastic than the succeeding ones, as it will then take a firmer hold on the " dead " lead coat over which it is placed, contributing a portion of its elasticity to that coat, and also cling more firmly to the hard drying coats which follow.

The body having been filled in may be set aside to harden, or if the smith is ready for it this is the best time for him to take it in hand, as any dents or burns that he may cause can now easily be remedied without spoiling the appearance of the vehicle. Later on, this is a matter of great difficulty, if not impossibility. Any bruises should be puttied; any parts which may happen to be burned must have the paint scraped off bare to the wood. Prime the bare spots, and putty and fill them to bring them forward the same as the general surface.

A material has recently been brought into the English market called "permanent wood filling," which is confidently recommended as effecting a saving in time, expense, and labour, and at the same time more effectually closing the pores of the wood than the ordinary filling now in use. This invention is due to a Polish exile named Piotrowski, who took refuge in America, and there introduced it about 1867, since which date it has found its way into the chief carriage factories in the United States. It is applied to the bare wood, one coat being given to bodies and two coats to carriage parts. This closes the pores, holds the grain immovably in its place, and is so permanent in its effect that neither exposure to dampness, nor atmospheric changes, nor the vibrations to which a carriage is so subject can affect the grain. The satisfaction which this material appears to give to the Americans, who pride themselves on the superiority of their carriage-painting, ought to induce our English coach-builders to inquire after it; for if all that we hear of it be correct, it must assuredly be a valuable acquisition to the paint shop.

In rubbing down use pumice-stone. It is best to begin on top and follow on down, so that the filling water may not run down on to any part that has been finished. Water should not stand for any length of time on the inside of the body; and when the rubbing is completed wash off clean outside and in, and dry with a chamois kept for the purpose.

The body, when dry, receives a staining coat, and is to be carefully sand-papered over, the corners cleaned out, and put on a coat or two coats of dark lead colour, made of tub lead, lampblack, raw oil, and a small quantity of sugar of lead, and reduced to a proper consistency with Japan gold size and turpentine. When dry scratch over the lead colour with fine sand-paper, which will make it appear of a lighter colour; we shall then be able to detect any low or sunk places by reason of the shadow. Putty up any imperfections with putty made of lead and varnish, and when dry face down with lump pumice and water. Follow with fine sand-paper, when the surface will be in a condition to receive the colour coats. Sometimes, after cleaning off, another coat of dark lead colour is laid on.

Analysing the foregoing, we find we have used—

1 priming coat of lead (or leather parts, 2 coats of black varnish
 instead).
2 thin coats lead colour, and stopped up.
5 coats of filling-up (8 coats on leather parts).
1 staining coat, rubbed down and cleaned off.
2 coats dark red colour, stopped up, and carefully rubbed down.
1 coat dark lead colour.
———
12 coats, and ready for colour.

So much for the body parts. To the carriage parts two
coats of priming are laid on, which are worked in the same
way as those applied to the body. All cavities are then
stopped with hard stopper, to which a little turpentine is
added in order to make it sand-paper easily. Two coats of
quick-drying lead colour are then applied to the wood parts.
The whole is then well sand-papered down, and the grain
should be found well filled and smooth. A thin coat of oil
lead colour is then laid on, and when dry sand-papered
down; any joints or open places between the tire and felloes
of the wheels are carefully puttied up with oil putty. The
carriage parts are then ready for colour. This time we have
applied—

2 coats of lead priming, stopped up.
2 coats of lead, thoroughly sand-papered.
1 coat (thin) of lead colour, sand-papered and puttied up.
———
5 coats, and ready for colour.

The colours are to be ground very fine, kept clean, and
spread on with the proper brushes. If the panels are to be
painted different from the other parts, lay on the black first,
for if any black falls on the panel colour it will occasion some
trouble by destroying the purity of a transparent colour. By
repeatedly turning the brush over while using it, there is less
liability to accidents of this kind.

The colouring of the body is finished as follows :—For

the upper quarters and roof grind ivory black in raw oil to a stiff consistency, add a little sugar of lead finely ground as a drier, and bring to the required consistency with black Japan and turpentine. Lay on two coats of this, and then two coats of black Japan, and rub down. Then face off the moulding, and give a thin coat of dead black, after which apply another coat of black Japan, and flat again. The whole should then be varnished with hard drying varnish, flatted down, and finished with a full coat of wearing body varnish. The varnish should have at least three days to dry; five or six would be better. The first coat of rubbing varnish may be applied thinner than the others, in order to avoid staining the colours.

The pencils used on mouldings should be large enough to take in the whole width at once, and let the colour run evenly along, avoiding laps or stoppages, except at the corners, where it cannot be helped. Avoid the use of turpentine in varnish if possible; but if the varnish be dark and heavy, sufficient turpentine added to make it flow evenly will not hurt it. The half elastic and fine bristle brushes are better for working heavy varnish than the sable or badger.

In varnishing a body begin on the roof, bringing the varnish to within 2 or 3 inches of the outer edges. Next, the inside of doors, &c., then the arch. When these are finished, start on the head rail on one side, lay the varnish on heavy, and follow quickly to the quarter. The edge on the roof, which was skipped before, is to be coated and finished with the outside, thus preventing a heavy edge. Continue round the body, finishing the boot last.

The frames and other loose pieces about a coach should be brought forward along with the body, and not left as is often done. The frames are most conveniently handled by a device similar to a swinging dressing-glass; a base and two uprights stoutly framed together, allowing space for the frame to swing. It is held in its position by two pointed iron pins, one fixed and the other movable. This is very convenient for varnishing, as the painter can examine his work by tilting it to any angle, and thus detect any pieces of dirt, &c.

If the body is to be lake in colour, the lake should be ground in raw oil, stiff, and reduced with turpentine and hard drying varnish. The same with dross black and Indian red. Over lakes and greens two coats of hard drying varnish should be applied, and one coat of finishing.

If the body is to be blue, mix ultramarine blue with one-half raw oil and turpentine, and bring it to a workable consistency by thinning with hard drying body varnish. Give the body two coats, and after each a slight flatting ; then give two more coats of the same with varnish added.

When Prussian blue is used, two coats are applied, and white is added, if necessary, to bring it to the required shade. The blues will dry sufficiently well when merely ground in raw oil, stiff, and reduced with turpentine, and it is better not to add a drier over blues ; only one coat of hard drying body varnish should be given, and one finishing coat.

In no case should the painter allow his oil colours to dry with a gloss. He must always flat them and give them the appearance of dead colour. This is particularly important, in case rough stuff or quick-drying colour is to be used over it.

The carriage parts are finished as follows :—Two coats of lead colour are first laid on, composed in the same way as those for the body before the colour is applied. Then stop all parts requiring it with hard stopper, a little reduced with urpentine to sand-paper easily. To the wood parts apply two coats of quick lead, composed of dry lead and lampblack ground in gold size and thinned with turpentine. Sand-paper down thoroughly, and the grain will be found smooth and well filled up. A thin coat of oil lead colour is then applied, and sand-papered down when dry ; and at this stage any open parts between the tire and felloe of the wheels, &c.,

should be again stopped up with oil putty. A coat of colour varnish follows, then a second, with more varnish added. The parts are then flatted and striped ; another light coat of clear varnish is given, and after being flatted down the fine lines are added, and the whole is finished with a good coat of wearing varnish.

The carriage parts are generally painted one or two tones lighter than the colour of the panels of the body, except where the panel colour is of a hue that will not admit of it. Certain shades of green, blue, and red may be used on panels, but would not, when made a tint or two lighter, be suitable for a carriage part. Dark brown, claret, and purple lake would not be open to this objection, because, to the majority of persons, they are colours which are pleasing to the eye, both in their deep and medium tones.

When the panels are to be painted green, blue, or red, and the painter wishes to carry these colours on to the carriage part, it is better to use them for striping only, and let the ground colour be black.

A carriage part painted black may be made to harmonise with any colour used on the body, as the striping colours can be selected so as to produce any desired effect. Brilliant striping can be brought out on dark colours only, while, if the ground colour be light, recourse must be had to dark striping colours to form a contrast. The carriage part should not detract from the appearance of the body ; that is, there should be sufficient contrast between the two to bring out the beauties of the body. A plainly finished body will appear to better advantage on a showy carriage, and a richly painted body on one that is not very ornate.

In striping the carriage parts, the bright colours should be used sparingly. A fine line placed on the face of the spokes and naves, and distributed over the inside carriage, would look far better than when each side of the spokes, the faces, the naves, and felloes, &c., are striped on both sides.

The coatings of varnish contribute largely to the durability as well as beauty of a carriage part. The ground and striping colours are shown in their purity only after they are varnished and have a good surface, and the test of wearing depends on the quantity and quality of varnish applied.

Every carriage part should have at least two coats of *clear* varnish. The first coat of varnish to be applied over the colour and varnish; the second, a good finishing coat, possessing body, and good wearing qualities. Ground pumice and water must be used to cut down the varnish, otherwise the finishing coat will be robbed of its beauty.

In laying on the finishing coat, avoid the extremes of putting it on too thick or too thin. Lay on a medium coat. A thin one will appear gritty and rough; and one too heavy will sink in and grow dim.

From the above description, it will be seen that painting a coach is a tedious operation, and one which consumes a great deal of time in its execution ; but, if well done, the result will certainly be very satisfactory. In no case should the painting be hurried, for by allowing each coat of paint or varnish sufficient time to dry its durability is insured.

A considerable amount of time is generally spent by the painter in work which does not really belong to him—that of mixing and grinding his colours. Where the muller and slab are used, they occasion a great deal of labour, and the tones of the colours are liable to be injured by the heat generated in the process ; and even where the hand mills are used, the process is by no means so cleanly as it ought to be. And under the heading of waste, this must always be a source of loss to the manufacturer, for the painter, for fear of not mixing up sufficient colour for his use, generally prepares too great a quantity, and as a rule, the surplus is waste, for it is no use to employ stale colours in painting vehicles, however well it may do in house painting.

What we want is to have the colours ready ground for the

painter's hand, and against this has been urged the objection, that the delicate colours would lose their purity, and all colours be more or less affected by it. That this is utterly fallacious is seen by the fact that paints and colours ready ground and prepared are the rule in America. The invention of the machinery, &c., for this purpose, is due to Mr. J. W. Masury, of New York, and he grinds pigments of the hardest description to the most impalpable fineness without injuring the tones of the most delicate ; and by a process of his own preserves them, so that the painter has nothing to do but to reduce them to the consistency he may require for the work in hand. He says they effect a saving of from 20 to 50 per cent. both of labour and material. It is difficult to understand why so valuable an invention is not more general in this country.

Irregularities in Varnish.

Varnish is subject to various changes after having been applied to a body or carriage part. It crawls, runs, enamels, pits, blotches, smokes or clouds over, and in the carriage parts gathers up and hangs in heavy beads along the centre of the spokes, &c.

These irregularities will happen at times with the very best varnish and the most skilled workmanship, and sur-rounded with everything necessary to insure a perfect job.

The only reason that can be assigned for it is atmospheric influence. These peculiarities have occupied a large portion of the time of the trade, and no other solution has been arrived at than the above.

The defects of varnishes should be divided into two classes : those which take place while in the workshop and those which show themselves after the vehicle has left the hands of the maker. The defects which show themselves in the varnish room are those of " spotting," " blooming," " pin-holing," " going off silky," " going in dead." Those which take place afterwards are "cracking," "blooming," "mud-spotting," and loss of surface, sometimes amounting to its almost total destruction.

The two classes should be considered separately; and assuming that the workmanship is of the best quality, the latter class of defects, with the exception of blooming, are in no way attributable to the varnish ; and blooming is caused by the atmosphere being overcharged with moisture, as would be the case before a storm, and it is soon remedied. Cracking will arise from too great an exposure to the sun, just as any other material will be damaged by unfair treatment. Mud-spotting will arise from using the carriage in muddy or slushy roads before the varnish is properly dry. The loss of surface will depend largely on the coachman, who, from ignorance or negligence, may rub down the panels of a carriage until its glossy surface entirely disappears ; and if the stable is contiguous to the coach-house this destruction will be assisted by the ammoniacal vapours arising from the manure, &c.

The other defects belong to the inherent nature of the varnish as at present manufactured, and admitting the secondary cause to be atmospheric influence, it is necessary to inquire why it is that varnish should be subject to such influence. According to the usual way of making varnishes, we know that various metallic salts and chemical compounds are used to increase their drying properties. All these will contain a certain definite amount of water, termed " water of crystallization." If deprived of this water they lose their crystalline form, but they acquire a tendency of again assuming it by attracting to themselves a proportionate amount of water when it is brought within their power. Now the heat employed in making varnishes is sufficient to expel this water ; but the presence of the salts is sooner or later detected, for when the varnish is applied to the work these salts absorb moisture from the atmosphere, and by becoming partially crystallized cause what is known as

G

" blooming," " spotting," and "pin-holing." The tendency to bloom will always remain, even after the varnish has hardened. But if any of these effects take place in the varnishing room, while the varnish is drying, it will be fatal to the appearance of the carriage.

To insure as near perfection as possible we want a substitute for these objectionable driers, which will not be subject to atmospheric influence.

CHAPTER XII.

ORNAMENTAL PAINTING.

MONOGRAMS.

At the present time nearly all possessors of carriages have their private marks painted on some part of the panels. These take the form of monograms, initial letters, crests, and heraldic bearings or coats of arms. The monogram is the commonest. For crests and coats of arms a duty is levied, from which monograms are free.

A few examples are subjoined. They can be multiplied to any extent; and designing monograms and initial letters would be excellent practice for the apprentice.

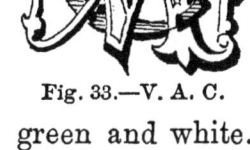

Fig. 33.—V. A. C.

Fig. 33.—Lay in C with dark blue, light blue, and chrome yellow, No. 2; lighted with A to be in Tuscan red, lighted with vermilion and orange; V with olive green, lighted with a bright tint of olive green and white. Separate the letters with a wash of asphaltum.

Fig. 34.—I. N. C.

Fig. 34.—Paint C a tan colour shaded with burnt sienna, shaded with asphaltum to form the darkest shades. Put in the high lights with white toned with burnt sienna. Colour I with dark and light shades of purple, lighted with pale orange; N to be lake colour lighted with

G 2

vermilion. The above may be varied by painting the upper half of the letters with the colours named, and the lower portions in dark tints of the same colour. When this is done, care must be taken to blend the two shades, otherwise it will look as if the letters are cut in two.

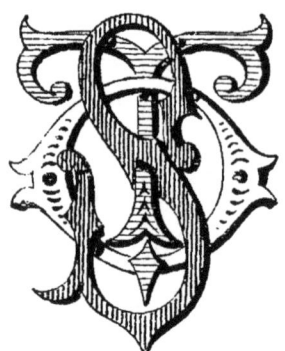

Fig. 35.—O. T. S.

Fig. 35.—Paint the upper half of O a light olive green, and the lower half a darker tone of the same colour; T to be lake, lighted with vermilion above the division made by the letter S, no high lighting to be used on the bottom portion of the stem; S to be painted red brown, lighted with orange ; or the colours may be laid on in gold leaf, and the above colours glazed over it.

Fig. 36.—This combination forms a pleasing variety, and

Fig. 36.—V. A. T.

will afford good practice in the use of the pencil. Lay in the letters as indicated by the shading, the letter V to be darker than A, and T deeper in tone than either V or A. The letters may all be laid on with gold leaf, and afterwards glazed with colours to suit the painter's taste. The vine at the base may be a delicate green tinged with carmine.

Fig. 37.—This is of French design. The letters furnish an odd yet attractive style. It will be noticed that the stem of the letter T covers the centre perpendicularly, and that the outer lower portions of A and R are drawn to touch on the same line. The main stems of these letters terminate in twin forms,

Fig. 37.—A. R. T.

arranged so as to cross each other at

the centre of the monogram and balance each other on either side. In the matter of its colouring, it may be mentioned that the letters in a monogram are very often painted all in one colour, and separated at the edges by a streak of white or high light. Monograms painted in this manner should be drawn so that the design will not be confused by ornamentation; that is, the main outlines of each letter should be distinctly defined, and the spaces must be so arranged as not to confuse the outlines. The pattern here given may be coloured carmine, and the edges separated by straw colour or blue, and the letters be defined by canary colour, or a lighter tint of blue than the bodies of the letters are painted.

Fig. 38.—T. O. M.

Fig. 38.—If the ground colour of the panels is claret or purple the letters may be painted with the same colour, lightened up with vermilion and white, forming three distinct tints; on brown, coat the letters with lighter shades of brown; and so on with other colours.

INITIAL LETTERS.

A well-painted initial letter is certainly quite equal to a monogram; but then it must be well painted, because, as it stands alone, it has only itself to rely upon for any effect, whereas, in a monogram, the component letters mutually assist each other.

Fig. 39.—This letter possesses all the grace of outline that could be desired in a single letter. Paint the letter in gold, shaded with asphaltum and lighted with white. If a colour be used,

Fig. 39.—D.

have one that agrees in tone with the striping on the carriage part; that is to say, if blue be used in striping, then use the same kind of blue for the letter, and so on with other colours.

We may here mention that all this kind of painting is done on the last rubbing coat of varnish, so that the letters receive a coat of varnish when the finishing coat is given.

Fig. 40.—The natural form of this letter is graceful, being composed of curves bearing in opposite directions, and which blend into each other, forming a continuous but varied line. The ornamentation also falls into the shape of the letter naturally. The upper and lower ends of the letter terminate in three stems, covered by three-lobed leafing, and the main stem of the letter is preserved in shape by appearing to grow out naturally from its outer and inner edges.

Fig. 40.—S.

Lay in the letter with gold, on which work out the design with transparent colours. If colours only be employed the panel colour may be taken as part of the colouring of the letter; for instance, if the panel be dark brown, lake, blue, or green, mix up lighter tints of whichever colour it may be, and considering the panel colour as the darkest shade, lighten up from it.

Fig. 41.—This letter will please by the novelty of its ornamentation. The body of the letter retains its natural outline almost wholly. From the upper part of the thin stem springs a scroll, which curves downward, reaching to the middle of the letter, and from this grows out a second scroll, serving to ornament the lower portions.

Fig. 41.—V.

Lay in the colour in harmony with the striping colour, deepening the tone of the colour on the stem of the letter, as shown by the shade lines. The leafing should be made out with light, medium, and dark tints, blended into each other so as to avoid the scratchy appearance which an opposite method produces.

CRESTS AND HERALDIC BEARINGS.

It would be impossible to give anything like a comprehensive series of these in this, or indeed in a very much larger work, as their number and variety are so great. The examples subjoined are given as exercises in colouring; and, if the student desires to extend his studies in this direction, most stationers will supply him with sheets of them at a trifling cost, and to them he may apply the principles enumerated below.

Fig. 42.—This is a small ornament, but it will disclose to the painter whether he has got hold of the method of handling the "cutting-up pencil." If, in attempting the circular part, the hand becomes inclined to be unsteady in its motion, and create a lack of confidence, the painter should practise until assured that the hand will obey the will.

Fig. 42.

The ornamental part to be gold, shaded with asphaltum, and high light with a delicate pink, composed of flake white and light red. The wreaths may be painted blue and white. Mix up three lines of blue, placing the darkest at the bottom or lower part of each band shown, as shaded in the figure. The white bands should not be of pure white, but a light grey, made by mixing a little black with the white colour. For the high light running along the centre of the wreath, use white tinted with yellow. The space covered with diagonal lines may either be left plain, showing the panel colour, or barred across with grey lines made of flake white and black, tinged with carmine.

Fig. 43.—This is the letter V combined with a garter. Size

in the entire pattern, and lay the pattern in with gold, and glaze over the inner part of the garter with a light blue, the inner and outer edges to remain gold. The flying ribbon to be pink, composed of carmine and white, and the shading to be clear carmine, with carmine saddened with black for the deeper tones. The

Fig. 43.

stems of the letter V to be green, shaded with a reddish brown, and the leafing to be the same colours.

Fig. 44.—Paint the cap crimson, the wreath green and grey,

lighted with a delicate pink. The circular part to be gold, shaded with asphaltum tinted with carmine; the outside border of shield to be gold also; the upper division of the shield to be red, deep and rich in tone. The chevron, or white angular band across the shield, to be a grey, lighted up with pure white. The lower division of the shield to be blue, and the deep shades to be purple. Paint the leafing at the base with a

Fig. 44.

colour mixed of burnt umber, yellow, and lake; shade with asphaltum tinted with carmine, and put in the high lights with orange or vermilion.

Fig. 45.—This is from a design by Gustave Doré. It is an
odd but still pretty design. Lay in the
whole of the pattern in gold ; shade
the details with verdigris darkened
with asphaltum ; put in the high
lights with pink, composed of light
red and white. The escutcheon
may be coloured with light brown,
carmine, and dark brown. The
edges of the diagonal bar to be dotted minutely with ver-
million.

Fig. 45.

Fig. 46.—Outline the garter with gold; the buckle and slide
to be gold also. Fill in the garter
with light and dark tints of blue, and
put in the high lights with canary
colour. Paint the floral gorgons
in brown shades, and light with
orange and clear yellow. A small
portion of lake added to these
browns will cause them to bear out
richly when varnished. Let the
medium lights and shades predomi-
nate, and the high lights added,
first carefully considering their true positions, and then
touching them in with sharp strokes of the pencil, which
will give life and " go " to the details. The pendent stems
with leaves and berries may be coloured olive green, and
shaded with russet. When the painting of this ornament is
dry it will be considerably improved by glazing.

Fig. 46.

Fig. 47.—The central pattern is *Caduceus*, a Roman emblem.
On the rod or centre staff the wings are represented " displayed," and the two serpents turning round it signify *power*,
the wings *fleetness*, and the serpents *wisdom*.

This pattern would look well in gold, with the dark parts
shaded with black to the depths shown on the sketch ; the

lighter tones being greys, warm in tone. The serpents may

be put in with carmine, as also the wings and head, and the rod carmine deepened with black.

Various treatments of colouring may be applied to this pattern, and thinking out some of these will be very good exercise for the ingenuity of the painter.

Fig. 48.—Put the pattern in in gold, separating the parts where necessary with shadow lines, and produce the effect of

Fig. 47.

interlacing by a judicious use of high light lines and deep

black lines. The best pencil suited to this class of ornament is a "cutting-up" pencil an inch long. Having traced the pattern on the panel, commence by painting the crest, and next the main upper left-hand division of the scroll part, paying no attention to the leafing or minor details. It will be noticed that the centre line of the heavy leafing is a part of the scroll line, which passes from the wreath or ribbon at the top, and is completed at the base; so that to secure easy curves this line should

Fig. 48.

be laid in through its whole length, and the leafing or any minor dividing lines be governed by it. Next lay in the other half of the pattern in the same manner, and having secured these main curves the subordinate details may be added.

Where two fine lines cross each other, the effect of one line passing underneath the other may be produced by simply lighting one of the lines across the intersection, which by contrast will make the gold or colour of the other line appear darker, and as though the lighted line passed over it and cast a shadow.

Paint the wreath blue and white, the crest to be merely lighted with the colour used for high lighting the other parts.

Fig. 49.—This consists of a species of dragon, having the head, neck, and wings of a bird, and the body of a wild beast. He supports a Norman shield, the "fess" or centre part displaying a Maltese cross.

Fig. 49.

In painting this ornament, first get a correct outline of the whole; then mix up two or three tints of the colour you design painting it, having a pencil for each, and a clean pencil for blending the edges, so that no hard lines may appear at the junction of the different colours. Lay on the shaded portions first, then the half lights, keeping them subdued in tone, so as to allow for the finishing touches showing clear and distinct.

On a claret-colour panel the whole may be painted in different hues of purple and red. On a dark blue panel, varying shades of blue lighter than the groundwork, and so with other colours. The shaded portion must be distinct, and gradually connected with the lighter portions by light tints of the shading colour.

Or the dragon may be painted grey, the high lights with the same colour warmed up with yellow; the outline of the shield in gold; the upper division, a light cobalt blue; the lower division, a pale orange; the cross, brown, shaded with asphaltum; the wreath, blue and white; and the flying ribbon and leafing in gold.

CHAPTER XIII.

LINING AND TRIMMING.

THIS is a department which requires great taste as well as skill. The interior of a carriage should be lined with cloth and silk, or cloth and morocco, with laces specially manufactured for the purpose. The colours should correspond to or harmonise with the painting. Light drab, or fawn colour, used to be a very general colour for the linings of close carriages, such as broughams, because they at once afforded relief to and harmonised with any dark colour that might have been selected for the painting. But a severe simplicity of taste has prevailed of late years in this country, and the linings of the carriages have been made mostly dark in colour to correspond to the colour of the painting. This is often carried to such an extreme as to present an appearance of sameness and tastelessness. It is no uncommon thing, for instance, to see a brougham painted dark green, striped with black lines, and lined with dark green cloth and morocco, with plain laces to correspond. This to us appears to be only one degree removed from a mourning coach, and it will be a great pity if such a taste prevails. On the other hand, violent contrasts outrage all principles of good taste. Morocco and cloth, or silk and cloth, of the same colour as the paint may be used for the linings, but, as the painting should be relieved by lines that harmonise with it, so should the linings be relieved by the laces and tufts, which are intended to give life and character to it.

Landau Back, Quarter, and Fall.

The back is made with one full row of squares, and two rows of buttons at the bottom, besides the finishing squares (see Fig. 50), then the swell of the back is carried up to within 4 inches of the upper edge, $1\frac{1}{8}$ inches being allowed for the swell. Then the top of the back is finished with a

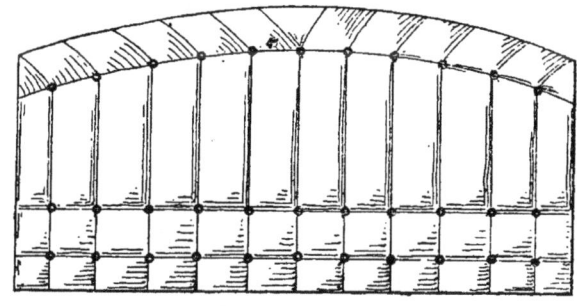

Fig. 50.

large roll, of about 5 inches swell or girth, so that the back has only one row of buttons in the upper sweep.

The arm-pieces, Fig. 51, are made in a peculiar way, and the *modus operandi* is rather difficult of explanation. In the place of the usual arm-piece block, a piece of plank, $2\frac{3}{4}$ inches wide by $\frac{1}{2}$ inch thick, is fitted in with the usual sweep to it. Now fit four pieces of single fly buckram, to form as it were a funnel, the shape of the arm-piece desired; then sew seaming lace to the two edges of the

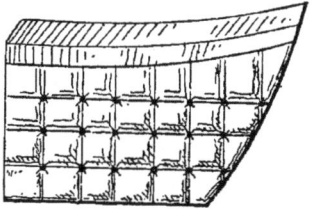

Fig. 51.

funnel, which will show inside of the body; to the lower edge sew a piece of cloth in smooth, so that it will cover the bottom of the funnel or cylinder. To the same lace edge, blind sew in a piece of cloth, for the purpose of forming a wrinkled roll on the inner face of the funnel, wrinkled 1 inch in fulness for every 3 inches in length, and as full the other way as desired, for it ought to be full enough to come out with the bottom side quarter. Next, blind sew the outer edge of this roll to the other or top seaming lace, and stuff lightly with hair, thus forming one roll on the inner face, and having the lower face covered with the smooth cloth.

Now blind sew another piece of cloth to the top lace, as in the other case, to form another wrinkled roll on top of the funnel or cylinder, but the outer edge of this roll is to be finished by nailing to the outside of the piece of plank just mentioned. All this sewing is of course to be done on the bench, one side of the funnel to be left open for this purpose. Next, nail in the bottom side quarter made up in squares, and then nail the side of the funnel, which is fitted against the arm-board, to the board and over the quarter, thus finishing the lower part. Then nail the fourth on top side to the top edge of arm-board; next, stuff from the front the funnel, pretty solid, and finish the top roll, which up to now has been left open, into the outside of the arm-board, thus completing the arm-piece, which shows two wrinkled rolls divided by two rows of seaming lace.

The door fall, Fig. 52, is made on three fly buckram pasted together, but one fly is cut off about an inch from the top to allow the fall to hinge. The fall is made about 12 inches deep, the lower edge being circular. The broad lace is bent to the required shape, and the corners sewed and put on to the buckram, which is cut to the shape intended for the fall. Mark where the inside edge of the lace comes all round the buckram, then mark 1¼ inches from the mark; paste a piece of carpet into the buckram to come within ½ inch from this mark, which will make the edge of the carpet 1¾ inches from the edge of the lace; cover this carpet with a piece of cloth, pasting on the buckram; take a piece of seaming lace, long enough to reach round the fall, and sew a piece of cloth to it, for the purpose of forming a wrinkled roll round three sides of

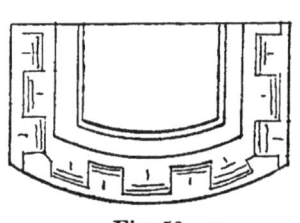

Fig. 52.

the fall, inside the broad lace and outside the sewed carpet, between both, with 1 inch fulness to each 3 inches of length ; sew this seaming lace and roll to the buckram at the mark, 1¼ inches from the broad lace, gather the other edge with a running string, and sew down and stuff lightly, finishing in such a manner that the broad lace shall cover this

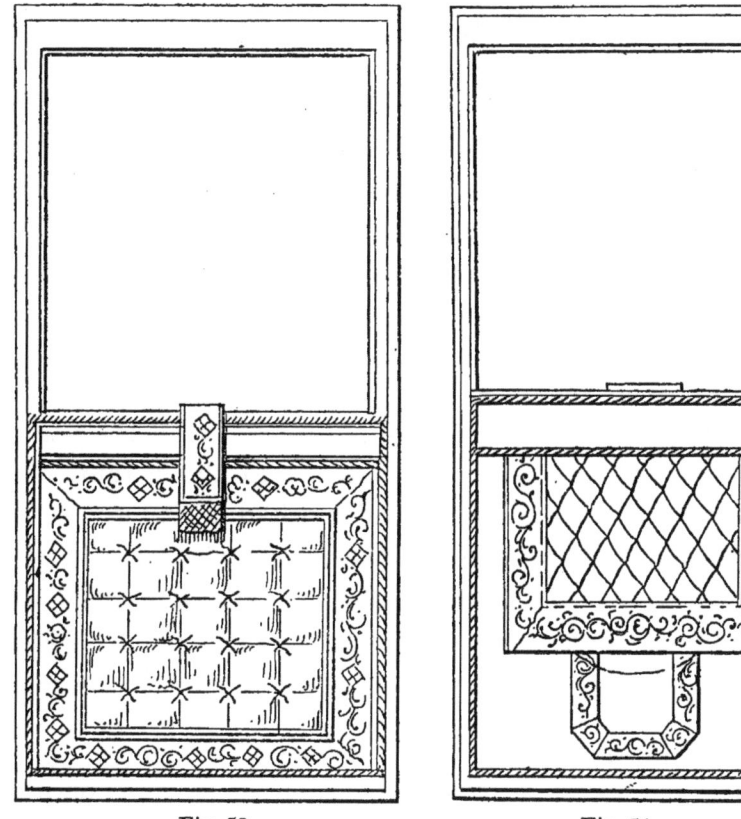

Fig. 53.   Fig 54.

sewed edge ; next, paste on the broad lace and cover the wrong side with silk or muslin. When dry, stitch both edges.

Figs. 53 and 54 show two styles of trimming for a door. Fig. 53 is made as follows :—Paste out three flies of buckram, and lay off for block or biscuit pattern, leaving space enough all round for a broad lace border, and at the top leave double the space. The top space is formed into a plain cloth roll of the same goods as the job is trimmed with.

In this case the trimming is brown cloth; the broad lace is silk and worsted of a shade much lighter than the cloth. The diamond-shaped and connecting figures are worsted and are raised.

The card-pocket is made of tin and covered with Turkey morocco the colour of the trimming.

In Fig. 54 it will be noticed that the style is somewhat different from the other; the surface of the door is trimmed plain, the fall alone being stuffed. The fall is stuffed in diamond form and enclosed with a lace border. This pocket runs up under the fall to the top, and is there nailed.

The following remarks on lining and trimming are taken from "Cassell's Technical Educator:"—

"We may with advantage say a word to our carriage lace-makers, who seem to have made but scant use of the various Schools of Design for the improvement of their taste in producing new and suitable patterns in the manufacture of their goods. For a long period we had nothing but the old scroll or flower pattern, which was handed down from father to son as if by a fixed law. At length, when it was felt that some change was required, the absence of all taste in design was shown in the production of entirely plain worked laces, which deprived carriage linings of their chief element of lightness and beauty. Thanks, however, to the taste and discernment of Messrs. Whittingham & Walker, who, perhaps, have devoted more attention to this branch of industry than any other house in London, the trade was relieved from the necessity of either adhering to the old pattern or of adopting the opposite scheme. They introduced small neat designs in laces eminently adapted to the purpose, and in 1857 they registered a pattern, now extensively known as the double diamond pattern, which has not only become general in England, but is largely patronised throughout Europe and America. This, and kindred patterns, exactly fulfil modern requirements, and give us the necessary relief without extreme.

" But with materials well and tastefully selected the trimmer has still his work to do. The lining of a carriage is divided into many different parts, all of which have to be designed. Canvas or paper patterns have to be cut to these, and properly fitted before the material is touched with which the carriage is to be lined. Wherever superiority of workmanship is to be shown in this department, the French method of trimming is adopted as being more elegant than the English. We shall therefore confine our observations to this method.

" In adopting the French method silk is mostly used in the place of morocco, and its peculiarities consist in the manner of quilting. The different squabs are made up in horizontal pipes or flutes, which are tufted in different ways. To proceed, cut a pattern in strong paper the size and shape of the space to be trimmed, and draw on it with a pencil the pipes, also mark the position of the tufts. On large pieces only mark one half, the other half being the same. The pipes of the back are usually 12 inches high and from 3 to 5 inches wide.

" The position of the tufts is considerably varied. Next stretch a piece of strong muslin in the stretching frame, lay the paper pattern on it, and mark the position of the tufts with an awl. Mark the lines of the pipes on the muslin with red chalk or pencil. In the same manner the pattern must be transferred upon the inside of the material used for covering, making of course due allowance for the depth of the pipes ; about 3 inches is a fair average for fulness at the top, 1 inch for the height of the pipes, and $1\frac{1}{2}$ inches for the width. For the last pipe an extra allowance is made in a narrow strip sewed on to it.

" Next lay a quantity of hair on the frame and form it the

swell desired.   Keep the hair in position with a few long
stitches, and lay the silk over it.   Commence tufting in the
middle of the lower row of pipes, and continue equally to
both sides.   Silk cords stretched into the channels between
the pipes were at one time considered elegant, but their
main merit was that they aided materially in preserving the
original shape of the pipes.   Backs are usually made couch-
shaped, with a roll all round on the top, which at the same
time form the elbow-pieces on the sides.   In elegant carriages
this roll is often elaborately executed in a helical or screw-
like shape, and continued from the door-pillar down to the
seat-frame, being made by winding silk cords around the
roll.   These silk cords appear as a single thread, but in
reality there are three different cords which are wound at
even distances.   A style of trimming much used of late both
in France and Germany for low backs, is a row of pipes at
the lower end, which are pinched to points at the top, and
above these are three rows of regular squares.   Squares are
preferred to diamonds as they are softer.

" Usually the back is laid on spiral springs, which are
fastened as follows :—The back of the body is covered with
coarse muslin, after being slightly stuffed, and on this muslin
four rows of seven small springs each are set.   For the lowest
row, springs a little stronger may be used than for the other
rows.   The highest row is set about $1\frac{1}{2}$ inches below the
edge of the back-board, and the lowest row at 6 inches
above the seat-frame.   The springs having been sewn on
with a bent needle, are tied first from right to left and then
from top to bottom.   A thin cord will answer for this
purpose.

" The cord is first cut in lengths, and when the tying
begins about 6 inches are allowed at the ends.   The cord is
wound about the third ring of the first and last spring in
each row, and afterwards the first ring is brought into the
right position with the piece of cord allowed over.   This

will make the spring stand upright, and it can be raised or lowered on one side. The springs being thus all placed in position, they are finally tied crossways.

· " The squab, in this instance, is worked in coarse muslin or canvas, stiffened with a little thin paste. It is set in the frame and marked as we have described above. When the cushions press against the back and side pieces, frequently no stuffing is made, but simply a piece of fine linen is sewed reverse to the main piece, and this is called the ' false finish.' In fine work the stuffing extends clear to the seat-frame.

" Of course each of these variations requires a different calculation for the muslin at the back as well as for the cover. For the latter an allowance of $1\frac{1}{2}$ inches is made for the pipes from the lower to the upper end, and also for the points an addition of $\frac{3}{4}$ inch. For each square in the height $1\frac{1}{4}$ inch has to be calculated. The folds of squares when laid over springs being diagonal, easily draw apart when stretched out, while the folds of diamonds running up and down may be drawn tighter to a certain degree of stretching.

" For the upper row of squares we have to allow for the backs at least double what we have to allow for the other rows—namely, 3 inches. For the width of every pipe an addition of $1\frac{1}{2}$ inches is calculated.

" Both cover and muslin being thus marked, we commence to draw in the tufts. Every point marked on the cover has to lie exactly on the corresponding one of the muslin. The lowest tufts are first drawn in ; then turn the frame and commence on one side at top, every point of the pipes being singly stuffed and the folds adjusted. This being done, every fold of the squares can be tufted right through, stuffed, and folded. Squares are easier to be worked than diamonds, but pointed pipes give more trouble than the ordinary straight ones.

" The elbow-pieces of this finish consist of two rolls made of muslin ; they are thinner towards the front of the seats.

After being stuffed, a piece of muslin 8 inches wide is sewed on all the length to the bottom of the roll, which serves, after the roll is tacked to the door-pillar and back, to give it

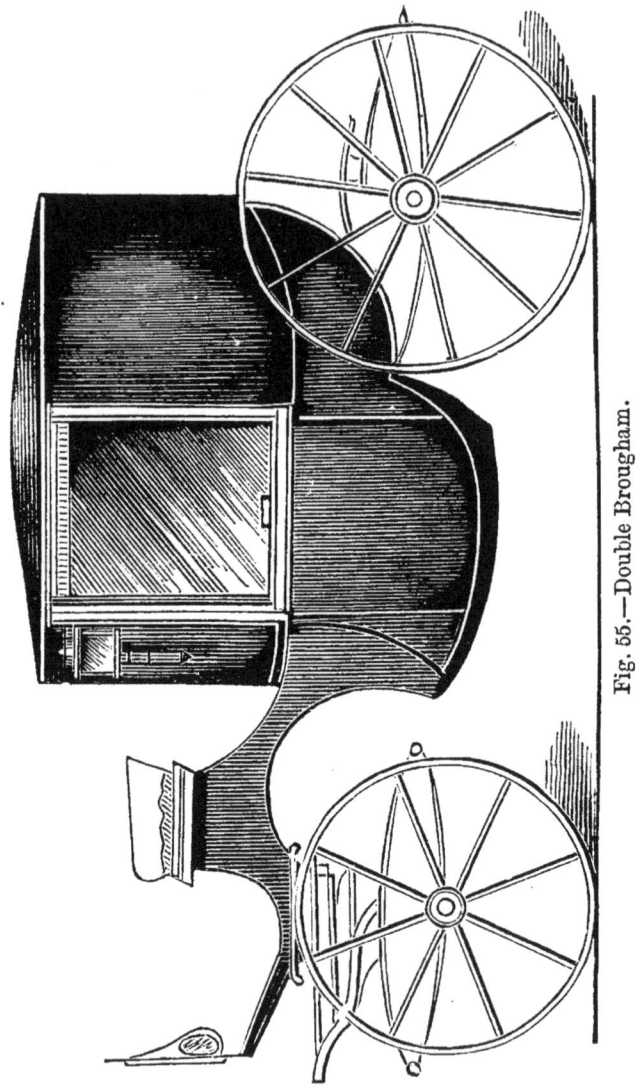

Fig. 55.—Double Brougham.

the required sweep in stretching and tacking it to the sides of the body.   Then mark on the roll the width of the pipes, and cut the cover for it, allowing 1 inch of width ; and as to the height, the cover must go all round, the roll having to be sewed back and front to the linen with which the roll is tacked to the body.

"After we have put in all the lining, we have to adjust the silk curtains, the blinds and glasses to doors and front part, to cover the iron dash-frame with the best patent leather, trim the coach-box seat, put on all the mouldings and bearings to the body, arrange the position of the lamps and fix them, and generally attend to all those little finishing points which give the appearance of neatness and finish to the whole."

Fig. 55 shows a double brougham with a circular front.

CHAPTER XIV.

GENERAL REMARKS ON THE COACH-BUILDING TRADE.

IT is pleasing to be able to refer to the increased skill and ingenuity of the coach workmen, especially among the rising generation of operatives. This fact was elicited by the recent exhibition of coaches, &c. Not only were there shown several excellent working drawings of carriages—drawn to scale and difficult of execution, and showing that there are forthcoming more highly educated and more competent men well acquainted with the details of their crafts, and of the proper and scientific manner of setting out their work, now that frequent change of construction renders this knowledge so desirable—but there were also shown many clever models of proposed improvements, the work of the ordinary carriage artisans, showing that their originators were men of thought and energy. And no doubt if more prizes were offered, and these exhibitions more frequent, greater competition would be aroused, which would be the means of bringing out a great deal of talent which at present lies dormant for want of some inducement to call it forth.

The art of the coachmaker being an intricate one, inasmuch as he has to combine in one harmonious whole a number of most varied products—wood, iron, steel, brass, paint, silver, cloth, leather, silk, ivory, hair, carpet, glass, &c., each worked by a separate trade, but generally in one manufactory, and each of which may be spoilt or injured by careless or improper treatment in any process—it behoves all engaged in

the production of carriages to work in harmony, that their
united labours may approach perfection. It would add much
to this desirable end if in each manufactory, large or small,
were issued a series of printed " general directions " for
conducting the work; not rigid rules that would, if strictly
enforced, reduce men to mere machines instead of free and
intelligent operatives, but such as would so guide each worker
in the execution of his work as not only to give satisfaction
to his employer by its excellent and honest execution, but
bring equal credit and satisfaction to himself. This state of
feeling would be a very desirable one to bring about; it would
be the means of bringing about mutual respect between employer and employed, and lead the way to a more cordial
appreciation of each other's wants and difficulties; at the
same time it would lessen the incessant watchfulness and
anxiety necessary to insure the work being executed in such
a manner that it may be depended upon for accuracy and
excellence when completed.

It is not so generally known as it should be, that in
France, Belgium, Germany, and some other European States,
the training of workmen and apprentices receives a great
deal of attention, the Governments in these countries considering money and trouble bestowed on such objects to be
of national importance. Technical schools in these countries
furnish instruction in drawing, modelling, the harmonious
arrangement of colours, the application of chemistry to
manufactures, metallurgy, and the proper working of metals,
the principles and applications of mathematics and mechanics
to manufactures, together with much that is strictly technical.
In some parts of Germany, before an employer of labour can
commence business on his own account, he must prove to
competent persons, by the execution of some trial work, that
he understands what he undertakes; and, moreover, that he
has travelled for three years in foreign countries, working at
his trade, to acquire a knowledge of its processes in other

countries besides his own. There is doubtless much pedantry in many of the regulations that interfere with the free exercise of trade, but culling the best points of the system there is much good that results. The training of apprentices in most trades in England is very unsatisfactory, and were public attention directed to the matter, after discussing the subject in its different bearings, there might be some good general recommendations circulated relating to the subject.

The carriages of America are so different from our own and from those of Europe, that they require special attention. It is quite possible that in the future their style may greatly influence carriages in all parts. The first noticeable trait in them is lightness, and English coach-builders generally agree that they carry this lightness too far, more especially in their larger carriages. We are supported in this view by the fact, that for some years, these—such as landaus, broughams, and coaches—have been materially modified by European types. The Americans have adopted some of the shapes of Europe, and the European mode of constructing the under-carriages, retaining their own method of making the pole and splinters, as giving greater freedom to the horses.

This principle of allowing the horses greater freedom for action is well worthy of the attention of coach-builders. The manner in which our horses are confined by tight, heavy strapping and traces, by tight pole chains, by bearing reins, and the indiscriminate use of blinkers to the bridles, has been much overdone in England. If a horse with a heavy load and driven fast over slippery roads should stumble, it is most difficult for him to recover himself. He generally falls, and is pushed along by the impetus of the carriage, and is more or less injured in his limbs or nerves by the accident, while it is a matter of great difficulty, if not impossibility, for him to rise again till the harness be unstrapped and the carriage is removed from above him. Our horses are also harnessed too closely to their work in two-wheeled carriages. We have thought only of the ease of turning and moving the vehicle in crowded or narrow ways, without observing the advantage of long shafts over short shafts. If the shafts are considered as levers, by which the horse supports and moves the weight behind him in a two-wheeled cart, it will at once be obvious that although (whilst those levers are parallel with the road) it does not so much signify whether they are long or short, yet the moment they cease to be parallel with the road, when they point upwards, or more particularly when they point downwards, the difference between long and short levers is severely felt by the horse. We can all of us lift a weight or support a weight more easily with a long lever than with a short one, and it is the same with a horse.

Those who have travelled abroad must have noticed the great weights placed upon two-wheeled carts in France and Belgium, and the greater comparative distance the horse is placed from the wheels, and yet he carries his load easily enough, because he does not feel its weight upon his back. Many English drivers seem to have observed this, and try and ease the horse and lessen his chance of stumbling by tipping the shafts up in front; but in this way the horse is made to feel a pressure on the under part of his body, which certainly will not improve his health. It is very probable that in future years public opinion will be in favour of longer shafts and poles. This will also tend to preserve good carriages from the damages they at present suffer from the heat of the horses and the quantity of mud which is thrown by their heels upon the front of the vehicle. The reins will of course have to be longer, but this cannot be of much consequence; the driver of a brougham is farther from the horse than the driver of a mail phaeton, but it is not by any means true that the brougham is any more difficult to drive than the phaeton on that account.

H

There is another fashion prevalent in this country which is certainly a fallacy, viz. the supposed necessity for the driver to sit nearly upright, which necessitates a deep boot and a clumsy, thick coachman's cushion. In America, Russia, and parts of Germany, the driver sits low, but places his foot against a bar in front of the foot-board; this in their carriages is longer than in ours. Four horses can be driven very well and easily in a low landau, and very powerful-pulling and fast-trotting horses held in with apparent ease. Our coachmen are often in danger of being pulled over by their horses, and certainly when an accident happens in a collision they are easily thrown from the boxes. They do not have the purchase and security that the Russian drivers seem to possess.

One of the greatest novelties introduced by the Americans into the United States is the "buggy," a name first given in England a hundred years ago to a light two-wheeled cart, carrying one person only, and which we now call a "sulky."

The Americans have lavished all their ingenuity upon these buggies, and they have arrived at a marvellous perfection of lightness. They are hung upon two elliptical springs. The axles and carriage timber have been reduced to mere thin sticks. The four wheels are made so slender as to resemble a spider's web. Instead of the circumference of the wheel being composed of a number of felloes, they consist of only two of oak or hickory wood, bent to the shape by steam. The ironwork is very slender and yet composed of many pieces, and in order to reduce the cost these pieces are mostly cast, not forged, of a sort of iron less brittle than our cast-iron. The bodies are of light work like what we call cabinet work. The weight of the whole vehicle is so small that one man can easily lift it upon its wheels again if it should be accidentally upset, and two persons of ordinary strength can raise it easily from the ground. The four wheels are of nearly the same height, and the body is suspended centrally between them. There are no futchells ; the pole or shafts are attached to the front axletree bed, and the front of the pole is carried by the horses in just the same way that they carry the shafts. The splinter-bar and whipple-trees are attracted to the pole on swivels. Some are made with hoods and some without. The hoods are made so that the leather of the sides can be taken off and rolled up, and the back leather rolled, removed, or fixed at the bottom, a few inches away from the back, the roof remaining as a sun-shade. The leather-work is very thin and of beautifully supple enamelled leather.

The perfection to which this vehicle has been carried is certainly wonderful ; and every part that is weak or likely to give way is carefully strengthened. If well made they last a long time without repair. The whole is so slender that it " gives " and recovers at any obstacle. The defect in these carriages in English eyes consists of the difficulty of getting in or out, by reason of the height of the front wheel and its proximity to the hind wheel. It is often necessary to partly lock round the wheel to allow of easy entrance. There is also a tremulous motion on a hard road that is not always agreeable. It is not surprising, however, that with the great advantages of extreme lightness, ease, and durability, and with lofty wheels, these vehicles travel with facility over very rough roads, as there is a great demand for them in our colonies. It must be remembered that the price is small, much less than the price of our gigs and four-wheeled dog-carts. This cheapness is attained by making large numbers to the same pattern, by the use of cast-iron clips, couplings, and stays, and by using machinery in sawing, shaping, grooving, and mortising the timbers, and by the educated dexterity of the American workman, always ready to adopt any improvement. An educated man will make a nimble workman, just as an educated man learns his drill from the military instructor more quickly than a clown ; and

an educated man finds out the value of machinery and desires
to use and improve it.　Instead of fearing its rivalry he wel-
comes it; he remembers that all tools, even the saw and the
hammer, are machines, and that the hand that guides these
tools is but a perfect machine obeying the guidance of the
brain more quickly and in a more varied manner than any
man-made machine.　The American workman, therefore,
uses machines more and more.

In England machinery for wood shaping is used at Derby,
Newcastle, Nottingham, Worcester, and other towns, and in
Paris some very good machinery is at work in coach factories.
In London it is chiefly confined to patent wheel factories, a
few steam-driven saws, patent mills worked by hand, and
drilling and punching machines.　But until the use of machinery is more generally adopted in London, it is probable
that the trade of building carriages for export will drift more
and more to the provinces and the continent.　The saving
effected by machinery in cab and omnibus building would be
great, because the patterns vary so little, and all the other
parts of a carriage would correspond with another, and
counter-change when repairs were needed.

The coach-builders of the future will look to steam and
hand machinery as their great assistance in cheapening the
cost of first-rate carriages, in multiplying them for the probable increased demand, and also to build carriages more
speedily.　It now takes from two to three months to build a
brougham, of which at least five weeks are consumed simply
in the wood and ironwork, a period which by the use of
machinery might easily be shortened.

There has been much controversy about difference in the
length of the front and hind axletree.　It has been usual to
make no greater difference than will allow the higher wheel
to follow in the same track as the lower wheel.　In France,
however, it has been the practice since the year 1846 to
make the front axletree of broughams 6 inches shorter than

the hind ones. The object has been to allow the front wheel to be placed nearer to the body. As the front wheel of a brougham must turn entirely in front of the body, the additional gain of 3 inches was very desirable. Some English coach-builders have followed the example of the French. There is a decided gain. The eye is pleased with the proportions, the horse is eased, and upon hard roads the difference of track is of no consequence. On the other hand, in the country roads, the well-worn ruts make the running of the carriage uneasy, whilst in town the driver often forgets that the curbstones will strike his hind wheels sooner than his front ones, and also more mud is thrown upon the panels. Under these circumstances it is very probable that the French plan will not find universal favour.

If carriages had always to move along perfectly smooth roads such as a tramway of wood, stone, or iron, the use of wheels in overcoming friction would be their sole utility, and their height would be of small consequence. But as carriages are drawn along roads with loose stones and uneven surfaces, wheels are further useful in mounting these obstacles, and it is plain that a high wheel does this more easily than a low wheel. To demonstrate this, let us suppose a shallow ditch or gulley of a foot wide and 2 inches deep, a wheel 2 feet high would sink into this and touch the bottom, but a wheel 3 feet high would only sink an inch, and a wheel 4 feet 6 inches high would only sink half an inch (the wheels are supposed to cross the above-mentioned gulley at right angles), on account of their greater diameters. Consequently, while the large wheel would have to be lifted by a force sufficient to raise it half an inch, a force will have to be applied to the smaller wheel to raise it 2 inches, and under more disadvantageous circumstances, because the spokes are in this case the levers, and we know that the longer the lever the more easily is the load raised.

That the leverage power of a high wheel is very great is

shown by the advantages gained by a large wheel in locomotives and bicycles.

There is an idea deeply rooted among coach-builders and coach-buyers too, that the draught of a vehicle is diminished by placing the front part of the carriage as far back as possible. Intelligent men who have given the subject great attention, and tested the actual working of this idea, say that it is a fallacy; but other intelligent men, who also say they have tested its working, say that it would effect a great saving in the draught if it were successfully accomplished. The general idea amongst practical men is, that it would not be an advantage. We have already seen that the draught of a vehicle with large wheels is less than that of a vehicle with small wheels. If, therefore, a load has to be placed on a four-wheeled vehicle, it should be so placed in relation to the front and hind wheels that the greater part of the weight should rest on the higher wheels. To obtain this result, it is sufficient to bring the hind-carriage part as far under the body as it will work with comfort and safety, in order that as little weight as possible rests on the fore-carriage part. English coachmakers have been working at this for thirty years, but for the most part blindly; they have copied well-known builders in construction as well as shape; they hear that these well-known firms' carriages run and follow very lightly, and if they could copy accurately they would obtain the same reputation.

But while there is some doubt as to throwing the front wheels backward for the purpose of lessening the draught, Mr. Offord of Wells Street, Oxford Street, has been exercising his ingenuity for the purpose of throwing the back wheels further forward, and has produced a brougham that offers peculiar advantages in this respect (Fig. 56). The hind wheel appears to be placed right across the door, but the facilities for ingress and egress are quite equal to those given in the ordinary brougham. This novel contrivance presents nothing singular in appearance, while a very little reflection will satisfy the practical thinker that the advantage sought after, of lessening the draught of the carriage, must be

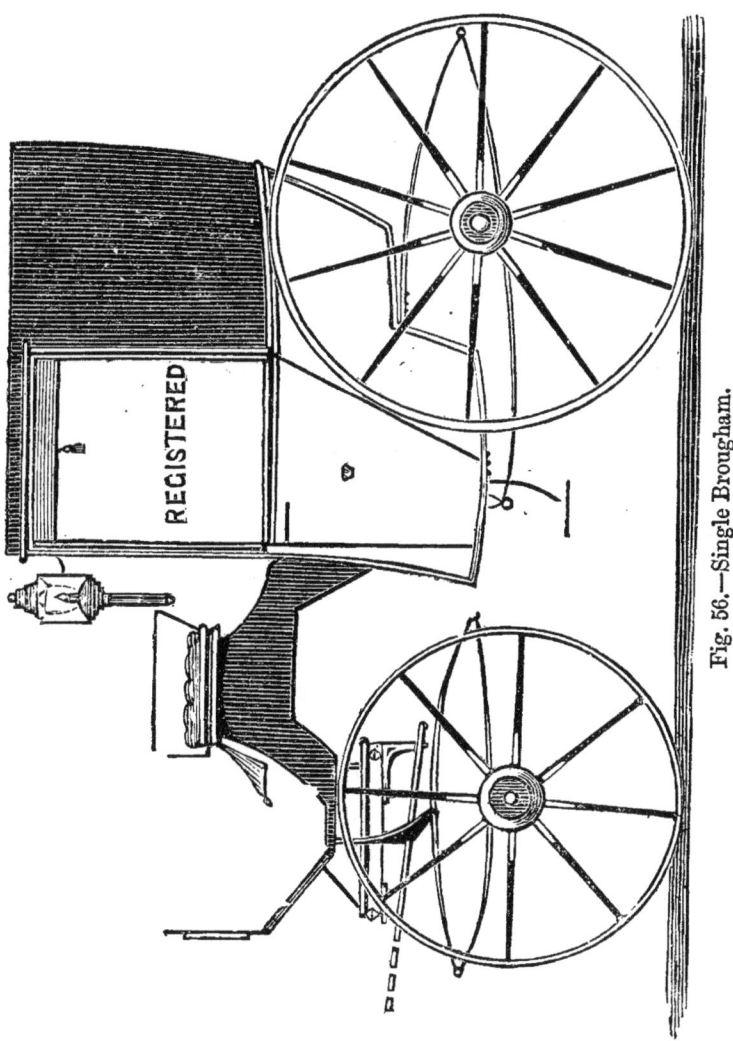

REGISTERED

Fig. 56.—Single Brougham.

obtained far more completely with this arrangement than by throwing the front wheel backward.

The rattling so constantly complained of in carriages can in a great degree be obviated by placing pieces of india-rubber so that the doors shall press upon them when closed; it is a good thing also to have india-rubber at the bottom of the doors for the windows to drop upon when let down.

The difficulty of protecting carriages from the dirt has recently been met by placing what is called a " mud scraper " just at the back of the hind wheels.   It is formed of a piece of india-rubber about 3 inches square and a $\frac{1}{4}$ inch thick, held in position by a short iron rod attached to the end of the hind spring.

The elaborate dress-carriages, hung with braces upon a C and underspring perch carriage, require so much skill and practice in their manufacture that it is impossible to give ample directions for their construction in a work like this. Some years ago it was thought that no carriage could be made comfortable unless it was hung upon the old-fashioned perch carriage with braces ; but it was found that by the intro- duction of india-rubber, especially in the ends of each spring, that what are termed elliptic spring carriages (which are of course much lighter in draught and less in cost) can be made extremely pleasant in motion.

Fig. 57 shows an iron-framed or "skeleton boot" for a landau.   It is extremely light and strong.

It is desirable to direct attention to the proper horsing of carriages, that the owners of carriages and horses may so adapt their plans as to get the most satisfactory result from their arrangements.   Not unfrequently a carriage is ordered for one horse only ; when it is partly made, or perhaps finished, fittings are ordered for two horses ; and it sometimes happens that the two horses put to the light one-horse carriage are coach-horses, between sixteen and seventeen hands in height. Such horses, though well adapted to a family carriage, are quite out of their place attached to a light one.   Although they can draw it at a good pace, and over almost any obstacle in the road, and do their journey without fatigue, the carriage suffers sooner or later.   The lounging of such horses against a light pole, the strain thrown on the pole in case of a horse tripping, the certain breakage that must occur in case of a fall, and the risk of overturning the carriage, should all be considered before putting a very light carriage behind very large horses. It also sometimes happens that miniature

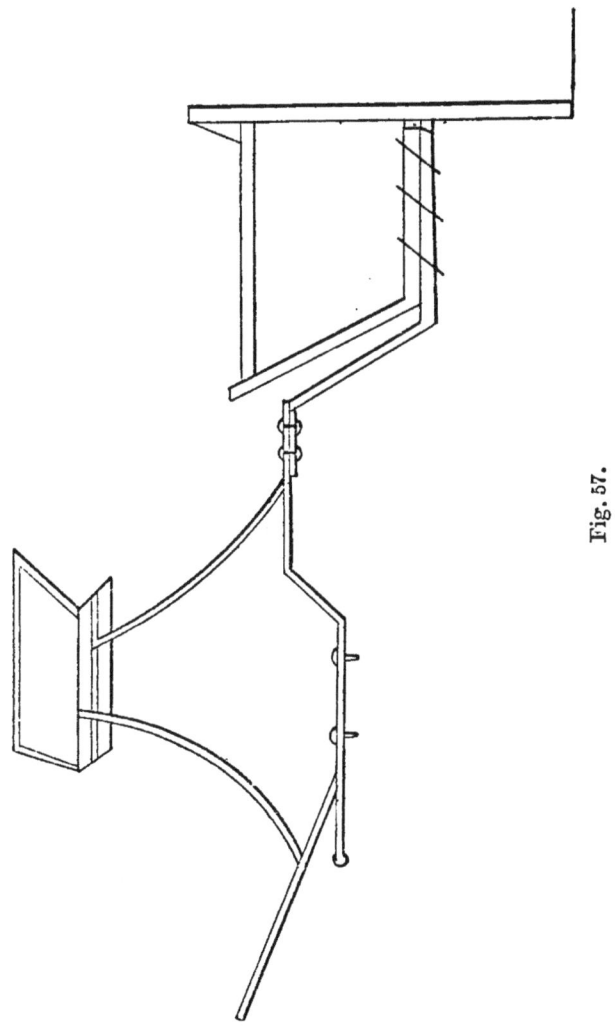

Fig. 57.

broughams and other very small carriages, built as light and as slight as safety will allow, are afterwards used with a pair of horses. In such cases, if accidents do not occur through the great strain of a long pole acting as a lever on very light mechanism, the parts become strained, do not work as they were intended to do, and necessitate constant repair from not being adapted to the work put upon them. Carriage owners should, in their own interest, have their carriages and horses suited to what they ought and can undergo, bearing in mind that there are advantages and disadvantages both with heavy and light carriages. The former are easier and more comfortable to ride in; they are safer for horses, drivers, and riders; and the necessary repairs are less frequently required. The lighter carriages follow the horses more easily, and can therefore do a longer day's journey; and, although the necessary repairs may come more frequently, the saving of the horses may be an advantage that many persons will consider of the utmost importance. Such light carriages should, however, be made of the choicest materials and workmanship, that they may do the work required of them.

A feature in the financial department of coach-making must not be overlooked, as it has much influence on an important trade. In former times a large proportion of the carriages were built to order for the owners; the reverse is now the case; most persons select a finished carriage which pleases their taste, or an advanced one, and get it completed their favourite colour. This, of course, necessitates the employment of a larger capital to meet the altered state of trade, which now requires so large a stock of carriages to be kept ready for use.

The excessive competition of recent years has so reduced the profit on each carriage, that in order to carry on his business without loss, the builder has to require a prompt payment from his customer instead of giving a long credit.

The modern system enables the coach-builder to make his purchases for ready money, and so buy not only better in quality but at a less cost than for extended credit, in order that he might in his turn give long credit to his customers, so that he is now obliged to depend on small profits and quick returns by turning over his capital more rapidly. He is not now, as much as in former times, the agent of the persons who supply the materials that he and his workmen convert into a carriage, but rather the designer, capitalist, and director of those who seek his service or custom, whether to supply labour or materials.

From the Government returns we find that carriages of all sorts have increased from 60,000, in 1814, to 432,600 in 1874—a benefit to the general population, it is clear, as well as to the workmen.   In 1874, 125,000 carriages paid the Government duty.

The valuable library and fine series of photographs of state and other carriages of the Coach and Coach Harness Makers' Company are open to coach artisans every Saturday afternoon.   Tickets of admission may be obtained at the principal coach-builders in London.

In Calcutta there are several coach-builders of good reputation, and who employ large numbers of native workmen.   Messrs. Dyke employ 600 hands ; Messrs. Stewart and Co., 400; and Messrs. Eastman, 300.   The men are chiefly Hindoos, and are clever and industrious, but have a singular habit of sitting down to their work.   Owing to the prejudices of the people in regard to the use of animal fat, the labourers who have to use grease are chiefly Mahommedans.   The wages in the trade vary from sixpence to two shillings per day.

"In Hindostan "(says Mr. Thrupp) "there are a large number of vehicles of native build.   It has been frequently remarked that there is little change in Eastern fashions, that tools and workmen are precisely as they were a thousand years ago, and the work they produce is precisely the same.   In examining, therefore, what is now done by Indian coach-builders, we are probably noticing carriages of a similar, if not identical, sort with those in use three thousand years ago.   The commonest cart in Hindostan is called 'hackery' by Europeans ; it is on two wheels, with a high axle-tree bed and a long platform, frequently made of two bamboos which join in front and form the pole, to which two oxen are yoked; the whole length is united by smaller pieces of bamboo, tied together, not nailed. In France two hundred years ago there was a similar cart, but the main beams terminated in front in shafts; in neither the cart of India nor of France were there any sides or ends. The French cart is called *haquet*, and it is probable that the French, who were in India as well as ourselves, may have given the term hackery to the native cart which was so like their own. The native name, however, is *gharry*. Other carts have sides made by stakes driven into the side beams; the wheels are sometimes of solid wood or even of stone. Wheels are also made by a plank with rounded ends and two felloes fitted on to complete the circle. Again wheels are made like ours, and also with six or eight spokes, which are placed in pairs, each pair close to and parallel with one another. If a carriage for the rich is required the underworks are like those of a cart, but the pole is carefully padded and ornamented with handsome cloths or velvet; the sides of the body are railed or carved, and the top is of a very ornamental character, similar to the howdah of state that is placed on an elephant. It has a domed roof supported upon four pillars, with curtains to the back and sides. The passengers ride cross-legged under the dome on pillows. The driver sits on the pole, which is broad at the butt end, and he is screened from the heat by a cloth which is fastened to the dome roof, and supported upon two stakes which point outwards from the body. A variety of different shaped native vehicles may be seen in elaborate models in the Indian Museum at South Kensington, although they do not show much originality of design or beauty of execution, and are said to be really creaking and lumbering affairs. When the Hindoos wish for a four-wheeled vehicle, the plan appears to be to hook on one two-wheeled carriage behind another, connecting them with a perch bolt, and on the hindermost they place the body. There is a singular addition to their vehicles outside the wheels; a piece of wood curved to the shape of the wheels is placed above it, frequently supported by two straight uprights from the end of the axle-tree outside the wheel. This acts as a wing or guard to keep any one from falling out of the vehicle, and also the dress of the passengers from becoming entangled in the wheel. In addition, a long bar of wood, rather longer than the diameter of the wheel, curved to the shape, called 'Cupid's bow,' is fastened to the axle-tree, the linch-pin being outside of it, and the ends of the bar tied to the ends of the wing by cords. I imagine it to be placed in order to be a safeguard for the people in crowded streets, who might be pushed by the throng against the wheel. It will be seen in many of the models, and also in ancient drawings of Indian and Persian vehicles. Many of the carts which are designed to carry heavy loads have a curved rest from 20 to 30 inches long attached to the lower side of the front end of the pole; this serves not only as a prop while the vehicle is being loaded, but should the oxen trip and fall it supports the cart and prevents the load, yoke, and harness from weighing down the poor animals, as they struggle to recover themselves. In England we have very few of these humane contrivances; we have, however, short rests to prop up a hansom cab when not at work. In India there are several huge unwieldy structures on wheels called 'idol cars;' the name of the car of Juggernaut must be familiar to many. The wheels of some of these are of enormous blocks of stone, shaped and drilled for the work. In the Indian Museum is a photograph of an idol car from South India, in the district of Chamoondee and the province of Mysore, which deserves examination. The car appears well proportioned, and the ornamental carvings are beautiful in design and would bear comparison with most European work.

"The *hecca*, or *heka*, is a one-horse native car, resembling an Irish car. It consists of a tray for the body fixed above the wheels on the shafts, and has a canopy roof; the driver sits on the front edge of the tray, and the passenger cross-legged behind him. The *shampony* is the usual vehicle for women, which resembles the former, but it is larger; the wheels are outside the body, and it is drawn by two bullocks; the canopy roof is furnished with curtains that are drawn all round, and the driver sits on the pole in front of the body. All these native vehicles have wooden axles, which until recently, I am told, were used without grease, from the prejudices of the people forbidding them to use animal fat. Some used olive oil or soap, but in most large towns there are now regulations obliging the natives to use some substance to avoid the noise and creaking of the dry axles. The commonest carriages in Central India are called ' *tongas*,' but the universal native word for a vehicle is ' *gharry*.' "

In 1860 a carriage was made for one of the ladies of the Sultan of Turkey's harem. It was built, I believe, from a design by the late Owen Jones—a great authority upon Oriental art—and cost £15,000 of English money—a very expensive present for the Commander of the Faithful to make to one out of many wives.

The following is given in an American trade periodical, under the heading of " Are they Competent Judges ? "—

" Carriage-makers who seldom if ever take the lines into their hands and ride out in carriages of their own manufacture, are they competent judges of the merits or the demerits of the vehicles which they with confidence recommend to others ? We think not. It is one thing to oversee and pay well for the building of a fine buggy or any other kind of vehicle, and quite another to experience the sensations produced by putting them into actual wear. A buggy may be handsome in general appearance and composed of the best material, yet defective in ease of motion and comfort to the occupants. The set of the axle may cause the vehicle to run heavy, and communicate to the rider an unpleasant jarring motion, and at the same time add unnecessary labour to the horse. The springs may be too stiff for their length, and fail to vibrate sufficiently under the greatest weight they may be called upon to sustain. The seat may be too low, the back placed in such a position or so trimmed as to be a continual source of uneasiness, and the foot-room be cramped. These and other defects may exist while the carriage-maker who seldom rides out remains in total ignorance of them, in so far as his own personal experience extends. Now an individual having purchased a buggy of such a one, might drive up to his door and inform him that this or that defect existed and needed to be remedied, and fail to convince the maker that such was the case. He would probably plead the skill of his workman, the care with which every buggy was carried forward to completion, and thus fortify himself in his own opinions, through gross ignorance of what constituted comfort while seated in a vehicle carried along over roads of different degrees of smoothness.

"The tendency of such a course is toward a standstill point in the way of needed improvement, and must certainly work adversely to the carriage-makers' interests. So far as our observation extends, we are well satisfied that the builder who adopts an opposite course is by far the most successful. Becoming sensible of defects by personal experience, he is keenly sensitive and anxious to remove any cause of complaint brought to his notice by others. With such a one the customer feels that he is dealing with a manufacturer alive to his convenience and comfort, and will not be apt to go elsewhere to purchase, although he may have had occasion to point out several weak points."

" The truly progressive carriage-maker tests his own work by frequently taking airing and criticisms of those who ride a great deal and are competent to speak on such points, and to any little defects that may be shown he gives the most careful examination and attention. No matter who may suggest a new idea of value, he puts it away as so much gained. He gathers here a little and there a little, which, in the aggregate, when applied as little things, amount to something so important as to give to his work an indescribable *something* which marks it as superior, and in short gives it a distinctive character."

It is too often the case that we look upon success in business as that condition only in which a man has secured to himself sufficient income to retire and lead a life of comparative ease and pleasure. While we would say nothing against an individual choosing to retire from active pursuits and enjoy the fruits of his labour, nevertheless the example thus set has a tendency to create in others a desire to *speedily* arrive at such a position careless of the means used to attain the end, and bringing into the business other elements than industry and the other good qualities necessary for the safe conduct of business, viz. grasping avarice, cunning deceit, and at times heartlessness, or, in fact, any legal means by which they can follow the American's advice to his son : " Get money, honestly if you can, but get money." These mean, sordid feelings of course react upon the employés, and they feel them in the shape of reduced wages and having the greatest amount of work literally ground out of them.

We hear occasionally of a man who, by a bold speculation, has " made a fortune " in a few months, but the majority of business men are not gifted with that keen foresight and courage which are so essential to the speculator, and must therefore be content with small gains, accumulating slowly year by year.

It is well that it is so, for the cares and disappointments attendant on the conducting of any business keep down pride of heart, and secure to society a majority of that class of men who can sympathise with the unfortunate and downtrodden, and who give more liberally to the rearing of those institutions which benefit and improve the masses.

Success depends in a great measure on the knowledge of the business engaged in, the proper application of industry to the materials required, frugality, promptness in meeting engagements, and good moral character.

In no occupation are the above qualities more essential than that of the manufacture of carriages, yet how few out of the whole number who claim to be carriage-makers have a good general knowledge of the business. Four distinct branches have to be looked after—woodwork, blacksmithing, painting, and trimming. The materials used by the respective branches are entirely dissimilar and costly, and require the utmost vigilance on the part of the proprietor to see that there is no unnecessary waste.

We shall close this chapter with the following remarks on " Taste " from W. Bridges Adams's valuable book on " English Pleasure Carriages : '—'

" There is a notion prevalent amongst uninstructed people that the quality called taste is a peculiar gift which an individual is endowed with at birth, and which cannot be acquired by any amount of application. Some portion of this belief is founded on reason, inasmuch as the physical faculties of some individuals at their birth are more perfect than those of others. Some are born with weak and some with strong eyes, and the same difference may exist in the perceptive faculties generally, on which faculties the quality of taste must depend. But even as weak eyes may be strengthened by judicious treatment, and strong eyes may be weakened by injudicious treatment, so inferior perceptive faculties may be improved by cultivation, and those which might have been first-rate may disappear by neglect. Even in those nations where the germs of taste are developed in but few individuals, where the mass of the community cannot discover beauty for themselves, they are yet susceptible of its influence when it is placed before them by others.

"Taste may be considered as another word for truth or proportion, both morally and physically. Much false taste exists in the community, and always has existed, but the total amount is continually lessening. The reason of the false taste is the imitative nature of man, which in an uncultivated state follows without examining. But even as it is the nature of water to attain a state of rest after violent oscillation, so it is the tendency of truth and proportion to grow out of the chaos of either thought or matter.

"Carriages constructed for the purposes of pleasure are works of art, in which taste may be widely developed in form, colour, and proportion, but the former is of course subservient to the mechanical construction. The hitherto defective mechanism of carriages, in which 'a large wheel is made to follow a small one,' has to a great extent destroyed proportion, and given a general license to all kinds of heterogeneous devices and barbarous ornaments, as if to overlay defects which there were no apparent means of obviating. Custom has reconciled the public to this discrepancy, which, were it now to appear for the first time, would excite universal distaste and ridicule.

"In an ordinary coach the side form of the body is composed of elliptic lines, from which the supporting iron brackets or loops are continued into reversed curves. This contrivance keeps the centre of gravity low. The four C-springs from which the body is suspended are each, or ought to be, two-thirds of a circle, with a tangent to it to form a base or support. The perch beneath the body, which connects together the framework supporting the spring, is curved into a serpentine line corresponding to the bottom of the body and the loops; and thus an agreeable form is preserved. But the double framework in front and the unequal wheels entirely discompose the whole effect, and from an art point of view it is extremely disproportioned, and consequently unsightly."

" Now the manufacturer possessing taste steps in, and by lightening the heavy parts by beading, carving, &c., fine lines of colour, and the arrangement of the hammercloth, redeems the vehicle from positive ugliness, and produces a work of art by the harmony of the various curves as a whole, though to produce this harmony there are no well-ascertained rules. Therefore it is that the builder who possesses taste produces combinations pleasing to the eye, and he who is without taste produces unsightly works, which he is necessarily obliged to sell at a low rate of profit as mere articles of convenience, not of refinement. And even as articles of convenience they are imperfect, inasmuch as the harmony of form arises from the due proportion of parts to each other, and that very proportion produces a greater amount of convenience. The size and weight of a carriage ought to be proportioned to that of the horse or horses intended to draw it, as well as the locality in which it is to be used and the persons who are likely to use it ; and the proportion of parts having once been accurately settled, the same rule of proportion must be observed, whether on an increasing or diminishing scale.

" After settling on the preliminary of form, the next consideration is that of colour. Taste in the latter can do much towards amending defects in the former, or at least can divert the attention of ordinary observers from dwelling upon them. Certain colours produce their effect by contrast, as green and red, purple and yellow, orange and blue, &c. ; others produce their effect by harmony, as green and drab, or brown and amber ; others again by gradation, as the differing shades of green and brown in almost endless variety. Colours are divided into two chief classes, the warm and the cold. Red and yellow and their varying gradations are warm colours. Green and blue and their varying gradations are cold colours. The intermingling of opposite colours produces neutrals. In choosing the colour for a carriage, it should be considered whether durability or appearance is the first consideration. For this country the warm colours are the most appropriate, as we hardly have enough of summer weather to render the adoption of cold colours general, except to such people as can afford carriages for each different season. The richest looking colours are not those which wear the best as a rule; but as an exception to this the yellows, which are both rich and showy, are amongst the most durable colours. For bright sunny days the straw, or sulphur yellow, is very brilliant and beautiful. Dark greens have a very rich appearance, but they do not wear well, the slightest specks being magnified by the dark surface. The olive greens are preferable, more especially for the summer, as they show the dust less, and are very good wearing colours. The shades of brown are even more numerous than those of the greens, and equally durable, though some of the lighter shades have a rather unpleasing effect, far too homely for varnish. Some of the darker browns become exceedingly rich with the admixture of a reddish tint, from the first faint tint up to the deep beautiful chocolate colour, the intermediate shades between which and a decided lake afford perhaps the very richest ground colours used in carriage-painting. Blues were formerly a great deal used to contrast with a red carriage part and framework. Very dark blues are now often used, but they soon become worn and faded, the least speck of dust disfiguring them. Drabs are scarcely ever used for body-painting, though for some peculiar purposes they might be advantageously applied.

" In addition to the ground colour other colours are used to relieve it, the framework of the body being generally painted black; and in the case of a very dark colour being used for the ground, it becomes necessary to run a fine line of a lighter shade in order to mark the inner edges of the framework. The same process is applied to the carriage parts and under framework for the purpose of making it look lighter to the eye. Were the perch, beds, and wheels painted of one colour they would look exceedingly heavy and clumsy; but the skilful management of the fine lines, or 'picking out,' as it is technically called, produces a pleasing optical illusion. The same effect is sought also in the carved work, which would look very bare were it not heightened and brought into relief by the judicious application of black and coloured lines. Heraldic bearings used to be painted very large on the panels; in fact they formed the principal ornament, as they were painted in their proper heraldic colours. With bright grounds, such as yellow, the effect is often very good, but with most other colours it destroys the general harmony, and on this account it has been the custom of late years to paint them very small, and very often of the same colour as the ground, only lightened up to give relief. This is of course the other extreme.

" Proportion in carriages applies to both form and colour; as regards form, it regulates the sizes of the various parts so that the whole may harmonise, and dictates the adoption of contrivances for lessening the apparent size of those parts which would otherwise be unseemly. Thus, the total height which is necessary in the body for the comfort of the passengers is too great for the length which it is convenient to give it; therefore the total height is reduced, and to give sufficient leg room a false bottom is affixed by means of convex rockers, and which, being thrown back and painted black, cease to form a portion of the elevation; they are, like a foundation, out of sight, and thus the proportion of the front view (the side is called the *front* in coach-builder's parlance) is preserved. In painting the body of a coach or chariot, it is customary to confine the ground colour to the lower panels and to paint the upper ones black, all except some stripes on the upper part of the doors. Now, inasmuch as colour in this case constitutes form by means of outline, and as that outline gives an irregular figure, it is a decidedly defective arrangement, making the upper part of the structure look heavier than the base. But the fact is, this defect has not been caused by intentional bad taste ; it is a mere result of imitation, of following up old practices when the motive for them has ceased. It was formerly the custom to cover the roofs and upper panels with greasy leather in order to make them water-tight, the edges of the leather being fastened down with rows of brass nails. This leather was black, and thus the eye became gradually reconciled to an unsightly object from a consideration of utility. After it was discovered that undressed leather could be strained on and painted, it was still considered necessary to paint it black, as the surface was not smooth enough to show well with bright colours ; and now that wooden panels are used to the upper as well as the lower part, long custom has made the black colour of the upper part appear indispensable.

" As by the present mode of constructing bodies various joints are left exposed to view, where leather unites with wood or two varieties of wood join in the same surface, it becomes necessary to resort to some means of covering them, and this is usually done by beading, as previously described. This is not altogether satisfactory as usually done, as it gives the side lines a broken and unfinished appearance. Where the beading is blacked it does not show much and scarcely matters, but the polished beading should go over the whole of the outline, as is done in some of the best carriages, or else it should not show at all. The elegance of a carriage depends on the perfection of the outlines, and anything which tends to disturb those outlines should be avoided.

" The handles of the doors are always made conspicuous, being of brass or plated metal. Necessity dictates this, as the constant action of the hand in opening or shutting the doors prohibits the use of paint on account of its rapid wear.

The side of the carriage would look better without this prominent projection if it could be avoided, but as that is impracticable it is generally placed at the intersection of the central vertical and the central horizontal lines, where it interferes less with the outlines than it would in any other position.

" In the lining and trimming of a carriage, form, colour, and proportion are all requisites. All dress carriages have hammercloths or coloured drapery surrounding the driver's seat. This forms a most prominent object, and if it does not harmonise with the rest of the vehicle the proportion of parts will be destroyed. The general form of the outline must be regulated by the lines of the ironwork or framework on which it is supported. There is great room for the display of taste in arranging that the colour of the hammercloth and lace, &c., shall harmonise or effectively contrast with the colour of the body. Yellow carriages are sometimes fitted with blue hammercloths and sometimes with drab ones, and the effect is equally good in both cases when well managed."

Careful attention to the above points will enable the practical coach-builder to produce a vehicle as near artistic perfection as the present shapes will allow.

CHAPTER XV.

INVENTION.

ENGLISH carriage constructors are certainly not an inventive race, if we allow that the names by which carriages are known are indicative of their origin. Coach is derived from the Hungarian *kotsee*, chariot is French, chaise is French, landau is German, cabriolet is French, and so on with many other names.

But mere invention—mere original conception—does not constitute excellence; and if foreigners may fairly lay claim to the greatest originality, English artists have the merit, perhaps still more important, of gradually improving the original designs, and so contriving all the details that, in their state of comparative excellence, the carriages can be scarcely recognised as constructions of the same principle as their models.

That English artists are not remarkable for the invention of new carriages is no proof of their want of talent; they have invention in abundance if there were sufficient motives to call it forth, and as a matter of fact invention is but poorly paid for. England possesses abundance of mind and matter, and there is no country in which a union of the two is just now more indispensable, and yet there is no country which throws greater obstacles in the way of the development of its minds. On an English patent lasting *fourteen* years the stamp duties amount to £175, whilst on an American patent lasting *seventeen* years the duty is only £7, or $\frac{1}{25}$th of the English stamp duty. Under such a system no one will be surprised to find that on December 31st, 1879, there were only 15,755 patents in force in England, as against more than 200,000 in the United States. (These figures, of course, refer to patents of all kinds.) It has been calculated that about 10 per cent. of patentees manage to survive the seventh year of their patents, at which time the £100 duty is payable; that is to say, as far as their patents are concerned.

English artists and artisans are little more than merchants in their trades and methods of doing business; they cannot afford to lose time, and their principal object is to make as large an annual return as possible, and as large a profit as possible on that return. Continental men are more enthusiastic lovers of their arts and sciences. They aim at improvement from mere liking for it, and when they fail it is mostly from want of efficient workmen to further their designs. The demand with them is not sufficient to make every branch of their art a manufacture. In England, on the contrary, the manufacture of carriages is a work of many trades, and greater skill is produced in manipulation by the division of labour. If chance brings in a new fashion, competition is aroused, which does not subside until some degree of improvement or excellence be obtained.

The ordinary measure of talent is held to be success, *i.e.* the acquisition of property; though it is quite clear the qualities which insure success are not those which tend to produce excellence or improvement in carriages more than in any other arts. The inventor may produce, but it is for the most part the mere merchant or tradesman who profits by the inventions. Carriages are made to sell as plays are written to fill theatres, and the English carriage-builder takes a French or German carriage to improve upon because it saves his time and trouble, just as the English play-writer freely uses a French play to save the labour of his brains. Improvements are rarely the voluntary productions of Eng-

lish carriage-builders; they are forced on them by the purchasers—first individuals and then the mass —who desire
some mere novelty, others greater ease, and others a more
rapid rate of motion. Almost all the changes and improvements in carriages may be traced in their origin to the carriage users and not the carriage-builders. The carriage-
builders do not lead, but they have always the means of
pressing talent enough into their service whenever a sufficient demand offers them a remunerating return. Coaches
were first invented on the Continent, but it was in England
that they were improved into public stages, capable of being
run 10 miles and upwards per hour for days and weeks
together.

This was not done at once, or by any one man: it was
the combined result of numberless small improvements,
forced on by the necessity of overcoming practical difficulties. Coach-builders have not been remarkable as a
scientific body. They have been, strictly speaking, "practical men;" and as the knowledge they have gained by
experience has not been carefully hoarded in books, carriage construction has remained a sort of occult matter,
without any specific theory attached to it. Each one, as he
is freshly initiated, gains his knowledge as best he can—from
verbal instruction or from a new series of experiments—and
thus a considerable portion of his time must elapse ere he can
have verified his judgment. Enough of this knowledge
exists in various brains which might suffice for the construction of a sound theory, but it would be a difficult operation to gather it together, for many petty feelings would be
at work.

Many experimentalists understand the word *theory* as
synonymous with falsehood or absurdity, as the very opposite of *practice*. It is clear that practice must be the ultimate verification of theory; but every true practice must
have a true theory belonging to it. The *theory* of a subject

is the *science* or philosophy of that subject; practice is the positive knowledge or proof of the soundness of the theory. But as theories are more plentiful than practices, and as many of them are not verified, there are of course many false ones. On this ground unscientific experimentalists have acquired the habit of regarding *all* theory as false, which is about as reasonable as it would be to assert that because falsehood exists in the world, all truth must therefore be extinct. This peculiarity is not confined to carriage-building; engineering and architecture abound with it; and law and medicine are not wanting in it. The truth is, that human knowledge is only got together by small portions at a time in the school of experiment, and when that knowledge is considerable in any one branch, a true and verified theory may be constructed from it. And when a great number of subjects have thus been analysed and theorised, it is comparatively easy to construct theories by analogy on new subjects by sound principles. Newton's theory of the universe was just as true when he first developed it in thought as after he had verified it by calculation.

It is a common notion that a mechanical inventor must necessarily be a man of genius; but, if the matter be analysed, it will be found that though inventors are occasionally men of genius it is not by any means a general rule. *Invention,* in its ordinary sense, as the word implies, is the art of *finding out.* By *genius* is meant a species of creative power, like that of the poet, for example, in his highest state of excellence. Invention is of two kinds—one resulting from a quick habit of observation, which detects the applicability of various forms of matter to similar objects. Of this an example may be given in the case of Dr. Wollaston, who, in a hurried experiment needing some lime which was not at hand, suddenly cast his eye on his ivory paper-cutter, and with some scrapings from its surface accomplished his object. This quick habit of observation, when it goes to the production of beautiful forms, is akin to fancy. The other and higher kind of invention is that which results from bringing a theory into practice—from first imagining a desirable result, and then bringing it to bear by the exercise of the judgment and constant persevering efforts steadily directed through a long period of time. The names of Brindley and Watt are examples of this quality. When Brindley set to work upon canals he did not create, he merely formed the plan of levelling the surfaces of natural streams by drawing off the water into new channels of sufficient depth, and thus preventing the water from being wasted. The process of forming locks was a continued series of mechanical contrivances with purpose aforethought. When Watt first imagined the steam-engine he did not invent the power of steam; that was known long before, and had existed from the time that fire and water had existed. But he formed to himself the plan by which he hoped to realise the result of making steam an efficient human servant through the agency of a perfect machine. The general idea of this machine existed in his mind a long time before he brought it into practice; and the slow process by which this was accomplished is evidenced by the fact that the term of his patent right was extended by Act of Parliament, on the ground that he had not had sufficient time to reap benefit from it. An anecdote of Watt serves not only to prove this, but also his high-minded philosophy, which was far beyond the miserable vanity of ordinary inventors, who aim at astonishing their fellows rather than instructing or benefiting them. After success had elevated Watt to the public eminence he so deserved, a nobleman who dined in his company expressed himself in terms of wonder on what Mr. Watt had accomplished. Watt coolly remarked, " The public look only on my success, and not on the intermediate failures and uncouth constructions which have served as steps to climb to the top of the ladder."

The power of mechanical conception is very widely extended; it is a modification of the same power which composes romances. The magic horse in the fairy tale, which turned and was guided by means of a pin in the neck, was a mechanical conception. If the same person who conceived that had worked it out into practice it would have been an evidence of genius, viz. imagination discovering truth by analogical inference. Invention, then, of the highest kind must be composed of four qualities—imagination, to conceive a new and complicated machine; knowledge, to gather materials; judgment, to select and combine them; and perseverance without wearying till the truth be obtained. Those things which commonly go by the name of new inventions are very often mere modifications of what has been done before. " Improvements " is the technical term applied to them.

The task a man has to go through in conceiving, designing, perfecting, and patenting a complicated mechanical invention is by no means inviting. Even when he possesses the hand to execute what the head has contrived, only a portion of his difficulties are overcome. A first idea is fascinating, and apparently easy of execution. It is thought over again and again; all difficulties are apparently surmounted and all obstacles removed; it is, in the imagination of the inventor, perfect. He may, perchance, know how to draw; but if not, he must employ some one to make his drawings for him. In this case, to avoid piracy, he must take out a provisional patent. For this he has to pay some considerable sum, and for a purpose whose success is uncertain.

Having secured his patent, our inventor sets his draughtsman to work and the drawings are made. Then follows the model, and ere that is completed it is discovered that there occur unexpected difficulties in the material construction which did not present themselves on paper. New contrivances must be resorted to, and the model is made and re-made many times over. It is at last completed, perhaps within a very few days of the time allowed for depositing the specifications, and fresh expenses are incurred by the necessity of paying highly those whose business it is to work against time. When all is ready the specification is deposited, and the inventor, perhaps, discovers that the title he has first taken will not cover his invention on account of its being different from his first contemplation. The title must, therefore, be altered, and the fees paid over again. He now sets to work to construct a full-sized sample of his invention, and all his patience is needed. At length the invention is in a state for practical trial. Up to this point all seems well; but practice soon discovers a defect, not in principle, perhaps, but in detail. A second experiment is made without success; and many more follow ere the invention is completed alike in principle and execution. Then begins the task of getting it before the public. Perhaps the inventor has been sanguine, and has attempted to introduce it to the public in an imperfect state, and the consequent failures have excited a prejudice unfavourable to his object. This prejudice has to be overcome by repeated and unceasing exertions, and at length, perhaps when half the period of the patent right has expired, the inventor begins to reap the fruits of his skill and industry. Public rumour is ever fond of exaggeration, and he is soon supposed to be realising a large fortune, though most likely he is only beginning to pay his expenses. Competition is then at work, and rivals who have been at no expense or trouble imitate his invention, or make just so much alteration in it as they think necessary to evade his patent right. He goes to law with the pirates, and then, perhaps, makes the discovery that his title or specification is imperfect, and that he has been labouring for years to bring to perfection an invention from which he can reap no more pecuniary advantage than if he had confined himself to an ordinary trade in which imitation alone was necessary.

It is evident, therefore, that a man who possesses a good

trade as a coachmaker has little inducement to embark in the perilous field of invention. His time is mostly taken up with his ordinary business, and unless under very peculiar circumstances he has little or no time to study improvements. Those who have too little business to fill up their time are interested in producing new things in order to attract public attention. First ideas very often originate with mechanics and artisans, who have not the means of putting them into practice, unless in comparatively trifling improvements. Established tradesmen generally consider it their interest to discourage such things, as interfering with their plans and giving them more trouble without extra profit. Thus, when under-springs were first adapted to carriages, it was prophesied that they would be the ruin of the coach-making trade, by making carriages too durable. When the streets were Macadamised, wheelwrights and coachmakers alike complained that it was destruction to their trade. It is the same in other things. The Manchester cotton-spinner, who has a mill and machinery already erected, does not feel very benevolently disposed towards an inventor who contrives new machinery of a better class, by which he can underwork him and take away his trade. It may be taken as a general rule that new inventions are viewed with jealousy by all established tradesmen, on the ground that they are an individual advantage, and not in the outset advantageous to the trade in general. Therefore, they keep them down all in their power, and when they succeed it is by the circumstance that they are valuable in themselves, and that the customers of the tradesmen insist on having them.

Sound mechanical knowledge is less necessary to the fashionable coachmaker than taste. Taste is the one requisite, without which he cannot thrive, and which therefore constitutes his real business qualification. Taste is exhibited in form, colour, and proportion, and having this he can employ other persons to fill up the details. The general mechanism

of carriages does not vary, and the mechanism serves as a
skeleton framework which may be clothed according to
fancy. Therefore, to produce what is commonly called a
new carriage is a work of composition, and not invention.
It is a combination of already existing parts to form a new
arrangement. The tasteful combiner may know nothing of
his wheels or axles, or their due proportion of strength, but
he has the wheelwright to take the responsibility for him.
He may know nothing of the construction of springs, but the
spring-maker is at hand ready to calculate the requisite
strength according to an estimated weight; and if the
weight should prove more than was expected it is easy to
apply an extra plate. He gives a general drawing of the
framework, and a skilful workman knows how to apportion
the scantling, and build it strongly together. A skilful
smith makes his ornamental ironwork to a given form, and
takes all the responsibility of understanding and duly working the metal. The employer directs the painter what
portions to put in colours and what in ground colours; what
to make conspicuous and what to hide; what to lighten by
lines and what to leave heavy. The preparation of the
colours and the laying of them on are the work of the painter
alone. The enterprising maker also directs the trimmer as
to the general effect of the lining, and arranges the harmony
of the colours; but the trimmer has to study the best mode
of performing his work. The braces and other leather work
are left to the skill of the workman, who is mostly left to
select his own materials and apportion their strength; and
the ornamental metal work is the province of the plater, who
is responsible for its wear. It is clear, however, that in
addition to taste, it is necessary that the carriage constructor
should know how to draw, in order to effectually direct
those whom he employs, and also to facilitate operations with
the purchasers who may employ him to build for them.

To be a complete carriage constructor a man ought to be

familiar with all the branches before alluded to. But there are few mechanics of such universal knowledge, and still rarer is it that they combine such knowledge with taste. Even it would be scarcely possible for a single individual to carry on a large business and do everything in his own factory. It would require a very large capital and very large premises, and also an extensive mercantile knowledge and skill—which last is based on qualities the direct opposite of those which nourish the faculty of taste. Mercantile skill depends on calculation ; taste is a combination of imagination and observation. There are three modes in which carriage-building on a large scale may be successfully conducted : first, by a single individual whose only business is to combine parts, and who employs tradesmen for every separate branch ; secondly, by a single individual, who employs responsible superintendents in every branch at high salaries; and, thirdly, by a combination of partners, one possessing taste, another mechanical knowledge, a third mercantile knowledge, and so on. This last mode would assuredly produce the most certain result, provided the partners possessed the necessary moral qualities to assure the absence of suspicion, jealousy, &c., amongst themselves. If these evil qualities existed they would destroy unanimity, and thus render the business unproductive by preventing efficient arrangements.

It is probable that at a future time the workmen themselves will enter into some such combination, but it must be after the lapse of many years, as the principle of caste must be first eradicated amongst them, which is at present so fruitful a source of jealousy amongst the different branches. Ere that takes place the increasing plenty of capital will probably induce many capitalists to invest money in carriage-building, as they now do in house-building, giving shares and salaries to men of undoubted skill and probity in order to insure efficiency and perseverance.

CHAPTER XVI.

REMARKS ON KEEPING CARRIAGES.

WITH very few exceptions, it is to be supposed that the greater number of those who can afford to indulge in the luxury of carriages are desirous of enjoying them on the most economical terms consistent with good taste, not merely as an economy of money, but also of time and convenience.

There are three ways of obtaining the use of carriages : 1. By hiring them for a short period, as a few days or weeks. 2. By taking them on lease for a term of years. 3. By purchasing them ready made or to order. The first two ways are now going out of date as a general rule. They are, of course, the most expensive. Any one requiring a conveyance for a few days, or a week or two, had better have a cab ; and as for taking them on lease, well, about four years' hire-money would purchase the vehicle outright. So, all things considered, it would appear the most economical and convenient to purchase the carriage to start with, and when it is no longer of any use there will at least be a second-hand carriage to dispose of.

As a rule, carriages are not built to order. The customer either chooses one from the stock, or selects one very nearly completed and has it finished to suit his own taste. This, of course, requires a very large capital to be invested in the business of a coach-builder, and, as competition has of late years greatly reduced the price of vehicles without a corresponding reduction in the cost of their production, the manufacturer naturally desires that his business should be as nearly as possible a ready-money one, otherwise he will have to do as many small, and even makers with a fair business, had to do on the introduction of this system, viz. shut up shop and take to something else to earn his living by.

People often marvel at the great cost of carriages, but when they have read of the numberless processes each vehicle has to go through there will be no longer food for surprise, but wonder that they do not cost more.

After a carriage is purchased a knowledge of how to preserve it from the various atmospheric and other influences, and how best to keep it in good order, is very necessary; for if great care is not exercised in the housing and cleaning of a vehicle its beauty will be utterly destroyed. In order to attain this knowledge it is requisite to remember of what the vehicle is composed—as wood, metal, leather, hair, cotton, silk, linen, paint, varnish, &c.

The ordinary atmospheric influences of our climate, sun, frost, dust, rain, and mud, all exercise a deteriorating influence on the vehicle. The general temperature most congenial to the durability of the carriage is that of the workshop in which it is constructed. In atmospheric air containing a certain amount of moisture, wood possesses a certain standard of bulk. If it be subjected to the influence of an atmosphere containing a greater amount of moisture it increases in bulk, or, as it is popularly termed, it swells; in a drier atmosphere it shrinks and is apt to crack. To resist these evil influences all the wood used in carriages is well covered with paint, the surface of which will resist moisture. If this operation is well and carefully done it is very successful, but woe betide slop-work, though in no trade is there so little room for a scamping workman to flourish. The result of bad painting is that moisture sooner or later finds its way into the wood and spoils the glossy appearance, and if it be placed in a very dry situation the panels will split, just as ships' decks would leak if not wetted several times a day during the heat of the sun. This might be applied in a modified degree to carriages, more especially to the wheels.

If due allowance be made for expansion and contraction, the metal-work of carriages, as springs, suffers very little from heat or cold, but moisture is apt to work a very destructive influence upon it, especially where the paint is worn away by friction. There the rust seizes hold of it and gradually insinuates itself beneath the whole covering of the paint, which strips off in flakes. Beneath the surfaces of the spring-plates also rust is continually working damage, and disfiguring the appearance with dirty brown lines of oxide of iron on the exterior. Brass and plated work also are considerably affected by damp.

Leather suffers greatly from heat and damp; but, like timber, more especially when subjected to alternations of heat and moisture. Toughness and tenacity are the chief qualities required for leather for carriages, and these qualities depend chiefly on the presence of a certain quantity of oil or fatty matter which the leather imbibes like a sponge. On this matter the oxygen of the air acts strongly, and at length consumes it; and if it be not renewed the leather cracks. If the leather be exposed to wet and damp this process is more rapid, but when the leather is frequently oiled it is apt to look dull and occasion much trouble to the coachman, who most likely will prefer blacking it, but the materials of which blacking is composed tend to the decomposition and destruction of the leather. Leather which is painted or japanned possesses little or no tenacity, and is never oiled. The patent grained elastic leather, which is so very much in use for hoods and knee-flaps, is a very beautiful substance to the eye, and is quite waterproof so long as it is free from cracks; but dryness and heat are liable to cause it to crack. Also, if one portion of the surface be kept in contact with another portion during warm weather, it is liable to stick and strip away when pulled apart. When it cracks and water gets in, it decays rapidly. Generally speaking, it is preferable to use oiled leather for heads, if ordinary care and attention be bestowed upon it ; for though its durability is not so great, there is a saving of labour in keeping it tidy, and it has a very good appearance.

The cloth, silk, and lace composing the lining, &c., and used in combination with wool, hair, cotton, and linen, suffer from the rays of the sun by losing their rich colours, and from the damp by becoming mildewed and rotten. Cloth, hair, and wool also suffer from another cause, viz. moths. In open carriages this is a very serious evil. Hammercloths are protected by a patent india-rubber cloth being put over them ; cedar shavings also exercise a destructive influence on moths. The india-rubber cloth is as good as anything where the smell is not objected to, but this in warm weather is very strong and unpleasant. However, it would be a very good thing to introduce some cedar shavings in the stuffing of linings, and this might to some extent get rid of the troublesome pest.

Simple damp does not cause much damage to paint and varnish unless it contains saline matter, then it is very destructive ; but heat, especially the strong rays of the sun, is very destructive. The colours change, and the lustre of the varnish disappears, and a multitude of intersecting cracks make their appearance; and to restore the original beauty there is no remedy but repainting. Another mischievous influence, acting on the paint and varnish, is the various gaseous vapours to which they are exposed. It is customary, for the sake of convenience, to stand carriages close to the horses' stables, generally in a mews, where large muck heaps are piled up in all stages of fermentation. During this process various gases are evolved, which act on the varnish just in the same manner that strong acids act on metals—by corroding or eating it away. The most destructive of these is the ammoniacal gas evolved from the urine.

It is evident that the ordinary coach-house is not the best that could be used for the purpose. The materials of a carriage are as delicate, and require as much care, as the furniture of a drawing-room, and therefore they should be as carefully preserved from stable contact as the satin couches of the drawing-room. After the carriage has been out, whether in the sun or rain, it should be carefully washed, and, above all, dried, taking care to wet the leather as little as possible during the operation. It is a common practice to wash the carriage and then leave the water to drip away. After drying, the leather should be carefully rubbed with an oiled rag, to restore the oily matter consumed by the vehicle being used. The carriage should then be placed to stand in a dry, well-ventilated apartment with a boarded floor, leaving a clear passage for the air beneath it, and if by any means convenient, let a current of warm air be passed through to insure its dryness. Above all it should be away from all stables, dung heaps, cesspools, or open drains. A gentleman should avoid placing his carriage in any situation where he would not wish to put his wardrobe; and with regard to the interior lining he should treat it in the same manner. If the carriage be laid by for a time it should occasionally be brushed out, and have a current of warm air passed through it. Cedar shavings should also be placed in it. If an open carriage it will require more care than a close one. The hammercloth (if there be one) should be covered with a waterproof india-rubber material, and cedar shavings interposed between the two. The blacking should also be rubbed off the leather work, and a composition of oil and tallow rubbed in to preserve it. The ironwork should be painted where any bare portions show themselves, caused by the rubbing of some other part against them.

Directions for keeping Carriages clean, &c.

Washing.—When a carriage is much used in the summer season use water freely, so as to remove dust or mud before using the sponge or chamois skin. The varnish of a carriage is often ruined through a want of attention to this matter, for the sharp particles of dust, which are chiefly silica, when by means of the leather forced over the surface of the varnish, act like diamonds on glass and score it in all directions. Mud should not be allowed to dry on the varnish if it can be avoided. The English varnishes take a long time to dry, and if mud gets on it before it is perfectly dry a permanent stain is left, which cannot be removed except by re-varnishing.

In winter time it is not a good thing to wash off the mud when it is so cold that the water freezes during the operation. *Warm water* should *never* be used in winter time, as it is apt to cause the varnish to crack and peel off.

Greasing.—For axles and wheel-plates the best lubricating material is castor oil. It is not necessary to apply a great deal at a time—little and often should be the rule; for when there is an excess of oil it oozes out and finds its way on to the stock, and from thence is thrown over the wheels while the vehicle is in motion. The grease is then liable to be taken up on to the sponge when washing, and also on to the leather, which will cause a great deal of trouble and vexation. The wheel-plate should be particularly looked after, and not allowed to become dry.

The Leather.—Enamelled leather should be kept soft and pliable with sweet oil or sperm oil. It will only be necessary, while the leather is new, to cleanse the top and curtains from dirt and rub them with a greased rag. When the leather shows signs of drawing up and becoming hard and lifeless, wash it with warm water and Castile soap, and with a stiff brush force the oil into the leather until all the pores are filled.

Sponges and Chamois.—Two of each of these should always be kept on hand, one of each for the body and the same for the under-carriage. The reason for this is, that after a carriage has been used there is a liability to get grease on the sponge and chamois after cleaning the wheels and wheel-plate mechanism. Another reason of some importance is that the sponges are soon destroyed by being used for cleaning the under-carriage, which renders them unfit for use for large panels.

The Cover.—When a vehicle has been washed and housed, it should be covered with an enamelled cloth cover, fitted to it so as to keep it free from dust inside and out. To preserve the wood and save expense it should be re-painted or varnished once a year. There is no economy in saving a few shillings this year if such saving will necessitate an expenditure of three times the amount next year.

INDEX.

K

THE END.

PRINTED BY J. S. VIRTUE AND CO., LIMITED, CITY ROAD, LONDON.

WEALE'S RUDIMENTARY SCIENTIFIC SERIES.

⁎ The volumes of this Series are freely Illustrated with Woodcuts, or otherwise, where requisite. Throughout the following List it must be understood that the books are bound in limp cloth, unless otherwise stated; *but the volumes marked with a ‡ may also be had strongly bound in cloth boards for 6d. extra.*

N.B.—In ordering from this List it is recommended, as a means of facilitating business and obviating error, to quote the numbers affixed to the volumes, as well as the titles and prices.

No.   ARCHITECTURE, BUILDING, ETC.

16. *ARCHITECTURE—ORDERS—*The Orders and their Æsthetic Principles. By W. H. LEEDS. Illustrated. 1s. 6d.

17. *ARCHITECTURE—STYLES—*The History and Description of the Styles of Architecture of Various Countries, from the Earliest to the Present Period. By T. TALBOT BURY, F.R.I.B.A., &c. Illustrated. 2s.
    ⁎ ORDERS AND STYLES OF ARCHITECTURE, *in One Vol.*, 3s. 6d.

18. *ARCHITECTURE—DESIGN—*The Principles of Design in Architecture, as deducible from Nature and exemplified in the Works of the Greek and Gothic Architects. By E. L. GARBETT, Architect. Illustrated. 2s. 6d.

⁎ *The three preceding Works, in One handsome Vol., half bound, entitled* "MODERN ARCHITECTURE," *price 6s.*

22. *THE ART OF BUILDING,* Rudiments of. General Principles of Construction, Materials used in Building, Strength and Use of Materials, Working Drawings, Specifications, and Estimates. By E. DOBSON, 2s.‡

23. *BRICKS AND TILES,* Rudimentary Treatise on the Manufacture of; containing an Outline of the Principles of Brickmaking. By EDW. DOBSON, M.R.I.B.A. With Additions by C. TOMLINSON, F.R.S. Illustrated, 3s.‡

25. *MASONRY AND STONECUTTING,* Rudimentary Treatise on; in which the Principles of Masonic Projection and their application to the Construction of Curved Wing-Walls, Domes, Oblique Bridges, and Roman and Gothic Vaulting, are concisely explained. By EDWARD DOBSON, M.R.I.B.A., &c. Illustrated with Plates and Diagrams. 2s. 6d.‡

44. *FOUNDATIONS AND CONCRETE WORKS,* a Rudimentary Treatise on; containing a Synopsis of the principal cases of Foundation Works, with the usual Modes of Treatment, and Practical Remarks on Footings, Planking, Sand, Concrete, Béton, Pile-driving, Caissons, and Cofferdams. By E. DOBSON, M.R.I.B.A., &c. Fourth Edition, revised by GEORGE DODD, C.E. Illustrated. 1s. 6d.

42. *COTTAGE BUILDING.* By C. BRUCE ALLEN, Architect. Ninth Edition, revised and enlarged. Numerous Illustrations. 1s. 6d.

45. *LIMES, CEMENTS, MORTARS, CONCRETES, MASTICS,* PLASTERING, &c. By G. R. BURNELL, C.E. Eleventh Edition. 1s. 6d.

57. *WARMING AND VENTILATION,* a Rudimentary Treatise on; being a concise Exposition of the General Principles of the Art of Warming and Ventilating Domestic and Public Buildings, Mines, Lighthouses, Ships, &c. By CHARLES TOMLINSON, F.R.S., &c. Illustrated. 3s.

83⁑. *CONSTRUCTION OF DOOR LOCKS.* Compiled from the Papers of A. C. HOBBS, Esq., of New York, and Edited by CHARLES TOMLINSON, F.R.S. To which is added, a Description of Fenby's Patent Locks, and a Note upon IRON SAFES by ROBERT MALLET, M.I.C.E. Illus. 2s. 6d.

111. *ARCHES, PIERS, BUTTRESSES, &c.:* Experimental Essays on the Principles of Construction in; made with a view to their being useful to the Practical Builder. By WILLIAM BLAND. Illustrated. 1s. 6d.

☞ *The ‡ indicates that these vols. may be had strongly bound at 6d. extra.*

LONDON : CROSBY LOCKWOOD AND CO.,

Architecture, Building, etc., *continued.*

116. *THE ACOUSTICS OF PUBLIC BUILDINGS;* or, The Principles of the Science of Sound applied to the purposes of the Architect and Builder. By T. ROGER SMITH, M.R.I.B.A., Architect. Illustrated. 1s. 6d.

124. *CONSTRUCTION OF ROOFS,* Treatise on the, as regards Carpentry and Joinery. Deduced from the Works of ROBISON, PRICE, and TREDGOLD. Illustrated. 1s. 6d.

127. *ARCHITECTURAL MODELLING IN PAPER,* the Art of. By T. A. RICHARDSON, Architect. Illustrated. 1s. 6d.

128. *VITRUVIUS—THE ARCHITECTURE OF MARCUS VITRUVIUS POLLO.* In Ten Books. Translated from the Latin by JOSEPH GWILT, F.S.A., F.R.A.S. With 23 Plates. 5s.

130. *GRECIAN ARCHITECTURE,* An Inquiry into the Principles of Beauty in; with an Historical View of the Rise and Progress of the Art in Greece. By the EARL OF ABERDEEN. 1s.
*** *The two preceding Works in One handsome Vol., half bound, entitled* "ANCIENT ARCHITECTURE," *price 6s.*
16, 17, 18, 128, *and* 130, *in One Vol., entitled* "ANCIENT AND MODERN ARCHITECTURE," *half bound,* 12s.

132. *DWELLING-HOUSES,* a Rudimentary Treatise on the Erection of. Illustrated by a Perspective View, Plans, Elevations, and Sections of a pair of Semi-detached Villas, with the Specification, Quantities, and Estimates, and every requisite detail, in sequence, for their Construction and Finishing. By S. H. BROOKS, Architect. New Edition, with Plates. 2s. 6d.‡

156. *QUANTITIES AND MEASUREMENTS,* How to Calculate and Take them in Bricklayers', Masons', Plasterers', Plumbers', Painters', Paperhangers', Gilders', Smiths', Carpenters', and Joiners' Work. By A. C. BEATON, Architect and Surveyor. New and Enlarged Edition. Illus. 1s. 6d.

175. *LOCKWOOD & CO.'S BUILDER'S AND CONTRACTOR'S* PRICE BOOK, for 1880, containing the latest Prices of all kinds of Builders' Materials and Labour, and of all Trades connected with Building: Lists of the Members of the Metropolitan Board of Works, of Districts, District Officers, and District Surveyors, and the Metropolitan Bye-laws. Edited by FRANCIS T. W. MILLER, Architect and Surveyor. 3s. 6d.; half bound, 4s.

182. *CARPENTRY AND JOINERY—*THE ELEMENTARY PRINCIPLES OF CARPENTRY. Chiefly composed from the Standard Work of THOMAS TREDGOLD, C.E. With Additions from the Works of the most Recent Authorities, and a TREATISE ON JOINERY by E. WYNDHAM TARN, M.A. Numerous Illustrations. 3s. 6d.‡

182*. *CARPENTRY AND JOINERY. ATLAS* of 35 Plates to accompany the foregoing book. With Descriptive Letterpress. 4to. 6s.; cloth boards, 7s. 6d.

187. *HINTS TO YOUNG ARCHITECTS.* By GEORGE WIGHTWICK. New, Revised, and enlarged Edition. By G. HUSKISSON GUILLAUME, Architect. With numerous Woodcuts. 3s. 6d.‡

188. *HOUSE PAINTING, GRAINING, MARBLING, AND SIGN WRITING:* A Practical Manual of, containing full information on the Processes of House-Painting, the Formation of Letters and Practice of Sign-Writing, the Principles of Decorative Art, a Course of Elementary Drawing for House-Painters, Writers, &c., &c. With 9 Coloured Plates of Woods and Marbles, and nearly 150 Wood Engravings. By ELLIS A. DAVIDSON. Third Edition, carefully revised. 5s. cloth limp; 6s. cloth boards.

189. *THE RUDIMENTS OF PRACTICAL BRICKLAYING.* In Six Sections: General Principles; Arch Drawing, Cutting, and Setting; Pointing; Paving, Tiling, Materials; Slating and Plastering; Practical Geometry, Mensuration, &c. By ADAM HAMMOND. Illustrated. 1s. 6d.

191. *PLUMBING.* A Text-Book to the Practice of the Art or Craft of the Plumber. With Chapters upon House Drainage, embodying the latest Improvements. Second Edition, enlarged. Containing 300 Illustrations. By W. P. BUCHAN, Sanitary Engineer. 3s. 6d.‡

☞ *The ‡ indicates that these vols. may be had strongly bound at 6d. extra.*

Architecture, Building, etc., *continued.*

192. *THE TIMBER IMPORTER'S, TIMBER MERCHANT'S,* and BUILDER'S STANDARD GUIDE; comprising copious and valuable Memoranda for the Retailer and Builder. By RICHARD E. GRANDY. Second Edition, Revised. 3s.‡

205. *THE ART OF LETTER PAINTING MADE EASY.* By J. G. BADENOCH. Illustrated with 12 full-page Engravings of Examples. 1s.

206. *A BOOK ON BUILDING, Civil and Ecclesiastical,* including CHURCH RESTORATION. With the Theory of Domes and the Great Pyramid, and Dimensions of many Churches and other Great Buildings. By Sir EDMUND BECKETT, Bart., LL.D., Q.C., F.R.A.S., Chancellor and Vicar-General of York. Second Edition, enlarged, 4s. 6d.‡ [*Just published.*

CIVIL ENGINEERING, ETC.

13. *CIVIL ENGINEERING,* the Rudiments of. By HENRY LAW, C.E., and GEORGE R. BURNELL, C.E. New Edition, much enlarged and thoroughly revised by D. KINNEAR CLARK, C.E. [*Nearly ready.*

29. *THE DRAINAGE OF DISTRICTS AND LANDS.* By G. DRYSDALE DEMPSEY, C.E. [*New Edition in preparation.*

30. *THE DRAINAGE OF TOWNS AND BUILDINGS.* By G. DRYSDALE DEMPSEY, C.E. New Edition. Illustrated. 2s. 6d.

31. *WELL-DIGGING, BORING, AND PUMP-WORK.* By JOHN GEORGE SWINDELL, A.R.I.B.A. New Edition, by G. R. BURNELL, C.E. 1s. 6d.

35. *THE BLASTING AND QUARRYING OF STONE,* for Building and other Purposes. With Remarks on the Blowing up of Bridges. By Gen. Sir JOHN BURGOYNE, Bart., K.C.B. Illustrated. 1s. 6d.

62. *RAILWAY CONSTRUCTION,* Elementary and Practical Instructions on the Science of. By Sir M. STEPHENSON, C.E. New Edition, by EDWARD NUGENT, C.E. With Statistics of the Capital, Dividends, and Working of Railways in the United Kingdom. By E. D. CHATTAWAY. 4s.

80*. *EMBANKING LANDS FROM THE SEA,* the Practice of. Treated as a Means of Profitable Employment for Capital. With Examples and Particulars of actual Embankments, and also Practical Remarks on the Repair of old Sea Walls. By JOHN WIGGINS, F.G.S. New Edition. 2s.

81. *WATER WORKS,* for the Supply of Cities and Towns. With a Description of the Principal Geological Formations of England as influencing Supplies of Water; and Details of Engines and Pumping Machinery for raising Water. By SAMUEL HUGHES, F.G.S., C.E. New Edition. 4s.‡

117. *SUBTERRANEOUS SURVEYING,* an Elementary and Practical Treatise on. By THOMAS FENWICK. Also the Method of Conducting Subterraneous Surveys without the Use of the Magnetic Needle, and other Modern Improvements. By THOMAS BAKER, C.E. Illustrated. 2s. 6d.‡

118. *CIVIL ENGINEERING IN NORTH AMERICA,* a Sketch of. By DAVID STEVENSON, F.R.S.E., &c. Plates and Diagrams. 3s.

197. *ROADS AND STREETS (THE CONSTRUCTION OF),* in two Parts: I. THE ART OF CONSTRUCTING COMMON ROADS, by HENRY LAW, C.E., revised and condensed by D. KINNEAR CLARK, C.E.; II. RECENT PRACTICE, including pavements of Stone, Wood, and Asphalte. Second Edition, revised, by D. K. CLARK, M.I.C.E. 4s. 6d.‡

203. *SANITARY WORK IN THE SMALLER TOWNS AND IN VILLAGES.* Comprising:—1. Some of the more Common Forms of Nuisance and their Remedies; 2. Drainage; 3. Water Supply. A useful book for Members of Local Boards and Rural Sanitary Authorities, Health Officers, Engineers, Surveyors, &c. By CHARLES SLAGG, A.I.C.E. 2s. 6d.‡

212. *THE CONSTRUCTION OF GAS-WORKS,* and the Manufacture and Distribution of Coal Gas. Originally written by SAMUEL HUGHES, C.E. Sixth Edition, re-written and much Enlarged by WILLIAM RICHARDS, C.E. With 72 Illustrations. 4s. 6d.‡ [*Just published.*

213. *PIONEER ENGINEERING.* A Treatise on the Engineering Operations connected with the Settlement of Waste Lands in New Countries. By EDWARD DOBSON, Assoc. Inst. C.E., Author of "The Art of Building," &c. 4s. 6d.‡ [*Just published.*

☞ *The ‡ indicates that these vols. may be had strongly bound at 6d. extra.*

MECHANICAL ENGINEERING, ETC.

33. *CRANES*, the Construction of, and other Machinery for Raising Heavy Bodies for the Erection of Buildings, and for Hoisting Goods. By JOSEPH GLYNN, F.R.S., &c. Illustrated. 1s. 6d.

34. *THE STEAM ENGINE*, a Rudimentary Treatise on. By Dr. LARDNER. Illustrated. 1s. 6d.

59. *STEAM BOILERS:* their Construction and Management. By R. ARMSTRONG, C.E. Illustrated. 1s. 6d.

67. *CLOCKS, WATCHES, AND BELLS*, a Rudimentary Treatise on. By Sir EDMUND BECKETT (late EDMUND BECKETT DENISON), LL.D., Q.C. A New, Revised, and considerably Enlarged Edition (the 6th), with very numerous Illustrations. 4s. 6d. cloth limp; 5s. 6d. cloth boards, gilt.

82. *THE POWER OF WATER*, as applied to drive Flour Mills, and to give motion to Turbines and other Hydrostatic Engines. By JOSEPH GLYNN, F.R.S., &c., New Edition, Illustrated. 2s.‡

98. *PRACTICAL MECHANISM*, the Elements of; and Machine Tools. By T. BAKER, C.E. With Remarks on Tools and Machinery, by J. NASMYTH, C.E. Plates. 2s. 6d.‡

114. *MACHINERY*, Elementary Principles of, in its Construction and Working. By C. D. ABEL, C.E. 1s. 6d.

139. *THE STEAM ENGINE*, a Treatise on the Mathematical Theory of, with Rules and Examples for Practical Men. By T. BAKER, C.E. 1s. 6d.

162. *THE BRASS FOUNDER'S MANUAL;* Instructions for Modelling, Pattern-Making, Moulding, Turning, Filing, Burnishing, Bronzing, &c. With copious Receipts, numerous Tables, and Notes on Prime Costs and Estimates. By WALTER GRAHAM. Illustrated. 2s.‡

164. *MODERN WORKSHOP PRACTICE*, as applied to Marine, Land, and Locomotive Engines, Floating Docks, Dredging Machines, Bridges, Cranes, Ship-building, &c., &c. By J. G. WINTON. Illustrated. 3s.‡

165. *IRON AND HEAT*, exhibiting the Principles concerned in the Construction of Iron Beams, Pillars, and Bridge Girders, and the Action of Heat in the Smelting Furnace. By J. ARMOUR, C.E. 2s. 6d.‡

166. *POWER IN MOTION:* Horse-Power, Toothed-Wheel Gearing, Long and Short Driving Bands, and Angular Forces. By J. ARMOUR, 2s. 6d.‡

167. *IRON BRIDGES, GIRDERS, ROOFS, AND OTHER* WORKS. By FRANCIS CAMPIN, C.E. 2s. 6d.‡

171. *THE WORKMAN'S MANUAL OF ENGINEERING* DRAWING. By JOHN MAXTON, Engineer. Fourth Edition. Illustrated with 7 Plates and nearly 350 Woodcuts. 3s. 6d.‡

190. *STEAM AND THE STEAM ENGINE*, Stationary and Portable. Being an extension of Mr. John Sewell's "Treatise on Steam." By D. K. CLARK, M.I.C.E. Second Edition, revised. 3s. 6d.‡

200. *FUEL*, its Combustion and Economy. By C. W. WILLIAMS, A.I.C.E. With extensive additions on Recent Practice in the Combustion and Economy of Fuel—Coal, Coke, Wood, Peat, Petroleum, &c.—by D. K. CLARK, M.I.C.E. 2nd Edition. 3s. 6d.‡ [*Just published.*]

202. *LOCOMOTIVE ENGINES*. By G. D. DEMPSEY, C.E.; with large additions treating of the Modern Locomotive, by D. KINNEAR CLARK, M.I.C.E. 3s.‡

211. *THE BOILERMAKER'S ASSISTANT* in Drawing, Templating, and Calculating Boiler and Tank Work, with Rules for the Evaporative Power and the Horse Power of Steam Boilers, and the Proportions of Safety-Valves; and useful Tables of Rivet Joints, of Circles, Weights of Metals, &c. By JOHN COURTNEY, Practical Boiler Maker. Edited by D. K. CLARK, C.E. 100 Illustrations. 2s. [*Just published.*]

216. *MATERIALS AND CONSTRUCTION;* A Theoretical and Practical Treatise on the Strains, Designing, and Erection of Works of Construction. By FRANCIS CAMPIN, C.E., Author of "Mechanical Engineering," &c., &c. 3s.‡ [*Just published.*]

217. *SEWING MACHINERY*, being a Practical Manual of the Sewing Machine; comprising its History and Details of its Construction, with full Technical Directions for the Adjusting of Sewing Machines. By J. W. URQUHART, C.E., Author of "Electro-Plating," &c., with numerous Illustrations. 2s.‡ [*Just published.*]

☞ *The ‡ indicates that these vols. may be had strongly bound at 6d. extra.*

SHIPBUILDING, NAVIGATION, MARINE ENGINEERING, ETC.

51. *NAVAL ARCHITECTURE*, the Rudiments of; or an Exposition of the Elementary Principles of the Science, and their Practical Application to Naval Construction. Compiled for the Use of Beginners. By JAMES PEAKE, School of Naval Architecture, H.M. Dockyard, Portsmouth. Fourth Edition, corrected, with Plates and Diagrams. 3s. 6d.‡

53*. *SHIPS FOR OCEAN AND RIVER SERVICE*, Elementary and Practical Principles of the Construction of. By HAKON A. SOMMER. FELDT, Surveyor of the Royal Norwegian Navy. With an Appendix. 1s. 6d.

53**. *AN ATLAS OF ENGRAVINGS* to Illustrate the above. Twelve large folding plates. Royal 4to, cloth. 7s. 6d.

54. *MASTING, MAST-MAKING, AND RIGGING OF SHIPS*, Rudimentary Treatise on. Also Tables of Spars, Rigging, Blocks; Chain, Wire, and Hemp Ropes, &c., relative to every class of vessels. With an Appendix of Dimensions of Masts and Yards of the Royal Navy. By ROBERT KIPPING, N.A. Fourteenth Edition. Illustrated. 2s.‡

54*. *IRON SHIP-BUILDING*. With Practical Examples and Details for the Use of Ship Owners and Ship Builders. By JOHN GRANTHAM, Consulting Engineer and Naval Architect. 5th Edition, with Additions. 4s.

54**. *AN ATLAS OF FORTY PLATES* to Illustrate the above. Fifth Edition. Including the latest Examples, such as H.M. Steam Frigates "Warrior," "Hercules," "Bellerophon;" H.M. Troop Ship "Serapis," Iron Floating Dock, &c., &c. 4to, boards. 38s.

55. *THE SAILOR'S SEA BOOK:* a Rudimentary Treatise on Navigation. Part I. How to Keep the Log and Work it off. Part II. On Finding the Latitude and Longitude. By JAMES GREENWOOD, B.A. To which are added, the Deviation and Error of the Compass; Great Circle Sailing; the International (Commercial) Code of Signals; the Rule of the Road at Sea; Rocket and Mortar Apparatus for Saving Life; the Law of Storms; and a Brief Dictionary of Sea Terms. With numerous Woodcuts and Coloured Plates of Flags. New, thoroughly revised and much enlarged edition. By W. H. ROSSER. 2s. 6d.‡ [*Just published.*

80. *MARINE ENGINES, AND STEAM VESSELS*, a Treatise on. Together with Practical Remarks on the Screw and Propelling Power, as used in the Royal and Merchant Navy. By ROBERT MURRAY, C.E., Engineer-Surveyor to the Board of Trade. With a Glossary of Technical Terms, and their Equivalents in French, German, and Spanish. Seventh Edition, revised and enlarged. Illustrated. 3s.‡

83bis. *THE FORMS OF SHIPS AND BOATS:* Hints, Experimentally Derived, on some of the Principles regulating Ship-building. By W. BLAND. Seventh Edition, revised, with numerous Illustrations and Models. 1s. 6d.

99. *NAVIGATION AND NAUTICAL ASTRONOMY*, in Theory and Practice. With Attempts to facilitate the Finding of the Time and the Longitude at Sea. By J. R. YOUNG, formerly Professor of Mathematics in Belfast College. Illustrated. 2s. 6d.

100*. *TABLES* intended to facilitate the Operations of Navigation and Nautical Astronomy, as an Accompaniment to the above Book. By J. R. YOUNG. 1s. 6d.

106. *SHIPS' ANCHORS*, a Treatise on. By G. COTSELL, N.A. 1s. 6d.

149. *SAILS AND SAIL-MAKING*, an Elementary Treatise on. With Draughting, and the Centre of Effort of the Sails. Also, Weights and Sizes of Ropes: Masting, Rigging, and Sails of Steam Vessels, &c., &c. Eleventh Edition, enlarged, with an Appendix. By ROBERT KIPPING, N.A., Sailmaker, Quayside, Newcastle. Illustrated. 2s. 6d.‡

155. *THE ENGINEER'S GUIDE TO THE ROYAL AND MERCANTILE NAVIES*. By a PRACTICAL ENGINEER. Revised by D. F. M'CARTHY, late of the Ordnance Survey. Office, Southampton. 3s.

55 & 204. *PRACTICAL NAVIGATION*. Consisting of The Sailor's Sea-Book. By JAMES GREENWOOD and W. H. ROSSER. Together with the requisite Mathematical and Nautical Tables for the Working of the Problems. By HENRY LAW, C.E., and J. R. YOUNG, formerly Professor of Mathematics in Belfast College. Illustrated with numerous Wood Engravings and Coloured Plates. 7s. Strongly half-bound in leather.

☞ *The ‡ indicates that these vols. may be had strongly bound at 6d. extra.*

LONDON : CROSBY LOCKWOOD AND CO.,

PHYSICAL SCIENCE, NATURAL PHILO-SOPHY, ETC.

1. *CHEMISTRY*, for the Use of Beginners. By Professor GEORGE FOWNES, F.R.S. With an Appendix on the Application of Chemistry to Agriculture. 1s.

2. *NATURAL PHILOSOPHY*, Introduction to the Study of; for the Use of Beginners. By C. TOMLINSON, Lecturer on Natural Science in King's College School, London. Woodcuts. 1s. 6d.

4. *MINERALOGY*, Rudiments of; a concise View of the Properties of Minerals. By A. RAMSAY, Jun. Woodcuts and Steel Plates. 3s.‡

6. *MECHANICS*, Rudimentary Treatise on; being a concise Exposition of the General Principles of Mechanical Science, and their Applications. By CHARLES TOMLINSON. Illustrated. 1s. 6d.

7. *ELECTRICITY;* showing the General Principles of Electrical Science, and the purposes to which it has been applied. By Sir W. SNOW HARRIS, F.R.S., &c. With Additions by R. SABINE, C.E., F.S.A. 1s. 6d.

7*. *GALVANISM*, Rudimentary Treatise on, and the General Principles of Animal and Voltaic Electricity. By Sir W. SNOW HARRIS. New Edition, with considerable Additions by ROBERT SABINE, C.E., F.S A. 1s. 6d.

8. *MAGNETISM;* being a concise Exposition of the General Principles of Magnetical Science, and the Purposes to which it has been applied. By Sir W. SNOW HARRIS. New Edition, revised and enlarged by H. M. NOAD, Ph.D., Vice-President of the Chemical Society, Author of "A Manual of Electricity," &c., &c. With 165 Woodcuts. 3s. 6d.‡

11. *THE ELECTRIC TELEGRAPH;* its History and Progress; with Descriptions of some of the Apparatus. By R. SABINE, C.E., F.S.A. 3s.

12. *PNEUMATICS*, for the Use of Beginners. By CHARLES TOMLINSON. Illustrated. 1s. 6d.

72. *MANUAL OF THE MOLLUSCA;* a Treatise on Recent and Fossil Shells. By Dr. S. P. WOODWARD, A.L.S. Fourth Edition. With Appendix by RALPH TATE, A.L.S., F.G.S. With numerous Plates and 300 Woodcuts. 6s. 6d. Cloth boards, 7s. 6d.

79**. *PHOTOGRAPHY*, Popular Treatise on; with a Description of the Stereoscope, &c. Translated from the French of D. VAN MONCKHOVEN, by W. H. THORNTHWAITE, Ph.D. Woodcuts. 1s. 6d.

96. *ASTRONOMY.* By the Rev. R. MAIN, M.A., F.R.S., &c. New Edition, with an Appendix on "Spectrum Analysis." Woodcuts. 1s. 6d.

97. *STATICS AND DYNAMICS*, the Principles and Practice of; embracing also a clear development of Hydrostatics, Hydrodynamics, and Central Forces. By T. BAKER, C.E. 1s. 6d.

138. *TELEGRAPH*, Handbook of the; a Manual of Telegraphy, Telegraph Clerks' Remembrancer, and Guide to Candidates for Employment in the Telegraph Service. By R. BOND. Fourth Edition, revised and enlarged: to which is appended, QUESTIONS on MAGNETISM, ELECTRICITY, and PRACTICAL TELEGRAPHY, for the Use of Students, by W. MCGREGOR, First Assistant Supnt., Indian Gov. Telegraphs. 3s.‡

143. *EXPERIMENTAL ESSAYS.* By CHARLES TOMLINSON. I. On the Motions of Camphor on Water. II. On the Motion of Camphor towards the Light. III. History of the Modern Theory of Dew. Woodcuts. 1s.

173. *PHYSICAL GEOLOGY*, partly based on Major-General PORTLOCK's "Rudiments of Geology." By RALPH TATE, A.L.S., &c. Woodcuts. 2s.

174. *HISTORICAL GEOLOGY*, partly based on Major-General PORTLOCK's "Rudiments." By RALPH TATE, A.L.S., &c. Woodcuts. 2s. 6d.

173 & 174. *RUDIMENTARY TREATISE ON GEOLOGY*, Physical and Historical. Partly based on Major-General PORTLOCK's "Rudiments of Geology." By RALPH TATE, A.L.S., F.G.S., &c. In One Volume. 4s. 6d.‡

183 & 184. *ANIMAL PHYSICS*, Handbook of. By Dr. LARDNER, D.C.L., formerly Professor of Natural Philosophy and Astronomy in University College, Lond. With 520 Illustrations. In One Vol. 7s. 6d., cloth boards.

*** Sold also in Two Parts, as follows :—

183. ANIMAL PHYSICS. By Dr. LARDNER. Part I., Chapters I.—VII. 4s.

184. ANIMAL PHYSICS. By Dr. LARDNER. Part II., Chapters VIII.—XVIII. 3s.

The ‡ indicates that these vols. may be had strongly bound at 6d. extra.

MINING, METALLURGY, ETC.

117. *SUBTERRANEOUS SURVEYING*, Elementary and Practical Treatise on, with and without the Magnetic Needle. By THOMAS FENWICK, Surveyor of Mines, and THOMAS BAKER, C.E. Illustrated. 2s. 6d.‡

133. *METALLURGY OF COPPER ;* an Introduction to the Methods of Seeking, Mining, and Assaying Copper, and Manufacturing its Alloys. By ROBERT H. LAMBORN, Ph.D. Woodcuts. 2s. 6d.‡

134. *METALLURGY OF SILVER AND LEAD.* A Description of the Ores; their Assay and Treatment, and valuable Constituents. By Dr. R. H. LAMBORN. Woodcuts. 2s. 6d.‡

135. *ELECTRO-METALLURGY;* Practically Treated. By ALEXANDER WATT, F.R.S.S.A. 7th Edition, revised, with important additions, including the Electro-Deposition of Nickel, &c. Woodcuts. 3s.‡

172. *MINING TOOLS*, Manual of. For the Use of Mine Managers, Agents, Students, &c. By WILLIAM MORGANS. 2s. 6d.‡

172*. *MINING TOOLS, ATLAS* of Engravings to Illustrate the above, containing 235 Illustrations, drawn to Scale, 4to. 4s. 6d. ; cloth boards, 6s.

176. *METALLURGY OF IRON.* Containing History of Iron Manufacture, Methods of Assay, and Analyses of Iron Ores, Processes of Manufacture of Iron and Steel, &c. By H. BAUERMAN, F.G.S. 4th Edition. 4s. 6d.‡

180. *COAL AND COAL MINING*, A Rudimentary Treatise on. By WARINGTON W. SMYTH, M.A., F.R.S. Fifth Edition, revised and enlarged. With numerous Illustrations. 3s. 6d.‡   [*Just published.*

195. *THE MINERAL SURVEYOR AND VALUER'S COMPLETE GUIDE*, with new Traverse Tables, and Descriptions of Improved Instruments ; also the Correct Principles of Laying out and Valuing Mineral Properties. By WILLIAM LINTERN, Mining and Civil Engineer. 3s. 6d.‡

214. *A TREATISE ON SLATE AND SLATE QUARRYING*, Scientific, Practical, and Commercial. By D. C. DAVIES, F.G.S., Mining Engineer, &c. With numerous Illustrations and Folding Plates. 3s.‡

215. *THE GOLDSMITH'S HANDBOOK*, containing full Instructions for the Alloying and Working of Gold, including the Art of Alloying, Melting, Reducing, Colouring, Collecting, and Refining ; Chemical and Physical Properties of Gold ; with a New System of Mixing its Alloys ; Solders, Enamels, &c. By GEORGE E. GEE, Goldsmith and Silversmith. Second Edition, considerably enlarged. 3s.‡   [*Just published.*

220. *THE SILVERSMITH'S HANDBOOK*, containing full Instructions for the Alloying and Working of Silver. By GEORGE E. GEE. 3s.‡

219. *MAGNETIC SURVEYING, AND ANGULAR SURVEYING*, with Records of the Peculiarities of Needle Disturbances. Compiled from the Results of carefully made Experiments. By WILLIAM LINTERN, Mining and Civil Engineer and Surveyor. 2s.   [*Just published.*

FINE ARTS.

20. *PERSPECTIVE FOR BEGINNERS.* Adapted to Young Students and Amateurs in Architecture, Painting, &c. By GEORGE PYNE. 2s.

40 & 41. *GLASS STAINING ;* or, The Art of Painting on Glass. From the German of Dr. GESSERT. With an Appendix on THE ART OF ENAMELLING, &c. ; together with THE ART OF PAINTING ON GLASS. From the German of EMANUEL OTTO FROMBERG. In One Volume. 2s. 6d.

69. *MUSIC*, A Rudimentary and Practical Treatise on. With numerous Examples. By CHARLES CHILD SPENCER. 2s. 6d.

71. *PIANOFORTE*, The Art of Playing the. With numerous Exercises & Lessons. From the Best Masters, by CHARLES CHILD SPENCER. 1s. 6 l.

181. *PAINTING POPULARLY EXPLAINED*, including Fresco, Oil, Mosaic, Water Colour, Water-Glass, Tempera, Encaustic, Miniature, Painting on Ivory, Vellum, Pottery, Enamel, Glass, &c. With Historical Sketches of the Progress of the Art by THOMAS JOHN GULLICK, assisted by JOHN TIMBS, F.S.A. Fourth Edition, revised and enlarged. 5s.‡

186. *A GRAMMAR OF COLOURING*, applied to Decorative Painting and the Arts. By GEORGE FIELD. New Edition, enlarged and adapted to the Use of the Ornamental Painter and Designer. By ELLIS A. DAVIDSON. With two new Coloured Diagrams, &c. 3s.‡

The ‡ indicates that these vols. may be had strongly bound at 6d. extra.

LONDON : CROSBY LOCKWOOD AND CO.,

AGRICULTURE, GARDENING, ETC.

29. *THE DRAINAGE OF DISTRICTS AND LANDS.* By
G. DRYSDALE DEMPSEY, C.E. *[New Edition in preparation.*

66. *CLAY LANDS AND LOAMY SOILS.* By Professor
DONALDSON. 1s.

131. *MILLER'S, MERCHANT'S, AND FARMER'S READY*
RECKONER, for ascertaining at sight the value of any quantity of Corn,
from One Bushel to One Hundred Quarters, at any given price, from £1 to
£5 per Qr. With approximate values of Millstones, Millwork, &c. 1s.

140. *SOILS, MANURES, AND CROPS.* (Vol. 1. OUTLINES OF
MODERN FARMING.) By R. SCOTT BURN. Woodcuts. 2s.

141. *FARMING & FARMING ECONOMY,* Notes, Historical and
Practical, on. (Vol. 2. OUTLINES OF MODERN FARMING.) By R. SCOTT BURN. 3s.

142. *STOCK; CATTLE, SHEEP, AND HORSES.* (Vol. 3.
OUTLINES OF MODERN FARMING.) By R. SCOTT BURN. Woodcuts. 2s. 6d.

145. *DAIRY, PIGS, AND POULTRY,* Management of the. By
R. SCOTT BURN. With Notes on the Diseases of Stock. (Vol. 4. OUTLINES
OF MODERN FARMING.) Woodcuts. 2s.

146. *UTILIZATION OF SEWAGE, IRRIGATION, AND*
RECLAMATION OF WASTE LAND. (Vol. 5. OUTLINES OF MODERN
FARMING.) By R. SCOTT BURN. Woodcuts. 2s. 6d.
*** Nos. 140-1-2-5-6, in One Vol., handsomely half-bound, entitled "OUTLINES OF
MODERN FARMING." By ROBERT SCOTT BURN. Price 12s.

177. *FRUIT TREES,* The Scientific and Profitable Culture of. From
the French of DU BREUIL. Revised by GEO. GLENNY. 187 Woodcuts. 3s. 6d.‡

198. *SHEEP; THE HISTORY, STRUCTURE, ECONOMY, AND*
DISEASES OF. By W. C. SPOONER, M.R.V.C., &c. Fourth Edition,
considerably enlarged; with numerous fine engravings, including some
specimens of New and Improved Breeds. 366 pp. 3s. 6d.‡

201. *KITCHEN GARDENING MADE EASY.* Showing how to
prepare and lay out the ground, the best means of cultivating every known
Vegetable and Herb, with cultural directions for the management of them all
the year round. By GEORGE M.F.GLENNY, Author of "Floriculture,"&c. 1s.6d.‡

207. *OUTLINES OF FARM MANAGEMENT, and the Organi-*
zation of Farm Labour: Treating of the General Work of the Farm; Field
and Live Stock; Details of Contract Work; Specialities of Labour; Economical Management of the Farmhouse and Cottage, and their Domestic
Animals. By ROBERT SCOTT BURN. 2s. 6d.‡ *[Just published.*

208. *OUTLINES OF LANDED ESTATES MANAGEMENT:*
Treating of the Varieties of Lands on the Estate; Peculiarities of its Farms;
Methods of Farming; the Setting-out of Farms and their Fields; the Con-
struction of Roads, Fences, Gates, and the various Farm Buildings; the
several Classes of Waste or Unproductive Lands; Irrigation; Drainage,
Plantation, &c. By R. SCOTT BURN. 2s. 6d.‡ *[Just published.*
*** Nos. 207 & 208 in One Vol., handsomely half-bound, entitled "OUTLINES OF
LANDED ESTATES AND FARM MANAGEMENT." By R. SCOTT BURN. Price 6s.

209. *THE TREE PLANTER AND PLANT PROPAGATOR:*
Being a Practical Manual on the Propagation of Forest Trees, Fruit Trees,
Flowering Shrubs, Flowering Plants, Pot-Herbs, &c.; with numerous Illus-
trations of Grafting, Layering, Budding, Cuttings, &c., Useful Implements,
Houses, Pits, &c. By SAMUEL WOOD. 2s.‡ *[Just published.*

210. *THE TREE PRUNER:* Being a Practical Manual on the
Pruning of Fruit Trees, including also their Training and Renovation, with
the Best Method of bringing Old and Worn-out Trees into a State of
Bearing; also treating of the Pruning of Shrubs, Climbers and Flowering
Plants. By SAMUEL WOOD. 2s.‡ *[Just published.*
*** Nos. 209 & 210 in One Vol., handsomely half-bound, entitled "THE TREE
PLANTER, PROPAGATOR AND PRUNER." By SAMUEL WOOD. Price 5s.

219. *THE HAY AND STRAW MEASURER:* Being New Tables
for the Use of Auctioneers, Valuers, Farmers, Hay and Straw-Dealers, &c.,
being a complete Calculator and Ready-Reckoner, especially adapted to
persons connected with Agriculture. Third Edition. By JOHN STEELE. 2s.
[Just published.

The ‡ *indicates that these vols. may be had strongly bound at 6d. extra.*

ARITHMETIC, GEOMETRY, MATHEMATICS, ETC.

32. *MATHEMATICAL INSTRUMENTS*, a Treatise on; in which their Construction and the Methods of Testing, Adjusting, and Using them are concisely Explained. By J. F. HEATHER, M.A., of the Royal Military Academy, Woolwich. Original Edition, in 1 vol., Illustrated. 1s. 6d.

*** *In ordering the above, be careful to say, " Original Edition " (No. 32), to distinguish it from the Enlarged Edition in 3 vols. (Nos. 168-9-70.)*

60. *LAND AND ENGINEERING SURVEYING*, a Treatise on; with all the Modern Improvements. Arranged for the Use of Schools and Private Students; also for Practical Land Surveyors and Engineers. By T. BAKER, C.E. New Edition, revised by EDWARD NUGENT, C.E. Illustrated with Plates and Diagrams. 2s.‡

61*. *READY RECKONER FOR THE ADMEASUREMENT OF* LAND. By ABRAHAM ARMAN, Schoolmaster, Thurleigh, Beds. To which is added a Table, showing the Price of Work, from 2s. 6d. to £1 per acre, and Tables for the Valuation of Land, from 1s. to £1,000 per acre, and from one pole to two thousand acres in extent, &c., &c. 1s. 6d.

76. *DESCRIPTIVE GEOMETRY*, an Elementary Treatise on; with a Theory of Shadows and of Perspective, extracted from the French of G. MONGE. To which is added, a description of the Principles and Practice of Isometrical Projection; the whole being intended as an introduction to the Application of Descriptive Geometry to various branches of the Arts. By J. F. HEATHER, M.A. Illustrated with 14 Plates. 2s.

178. *PRACTICAL PLANE GEOMETRY:* giving the Simplest Modes of Constructing Figures contained in one Plane and Geometrical Construction of the Ground. By J. F. HEATHER, M.A. With 215 Woodcuts. 2s.

179. *PROJECTION :* Orthographic, Topographic, and Perspective: giving the various Modes of Delineating Solid Forms by Constructions on a Single Plane Surface. By J. F. HEATHER, M.A. [*In preparation.*

*** *The above three volumes will form a* COMPLETE ELEMENTARY COURSE OF MATHEMATICAL DRAWING.

83. *COMMERCIAL BOOK-KEEPING.* With Commercial Phrases and Forms in English, French, Italian, and German. By JAMES HADDON, M.A., Arithmetical Master of King's College School, London. 1s. 6d.

84. *ARITHMETIC*, a Rudimentary Treatise on: with full Explanations of its Theoretical Principles, and numerous Examples for Practice. For the Use of Schools and for Self-Instruction. By J. R. YOUNG, late Professor of Mathematics in Belfast College. New Edition, with Index. 1s. 6d.

84*. A KEY to the above, containing Solutions in full to the Exercises, together with Comments, Explanations, and Improved Processes, for the Use of Teachers and Unassisted Learners. By J. R. YOUNG. 1s. 6d.

85. *EQUATIONAL ARITHMETIC*, applied to Questions of Interest,
85*. Annuities, Life Assurance, and General Commerce; with various Tables by which all Calculations may be greatly facilitated. By W. HIPSLEY. 2s.

86. *ALGEBRA*, the Elements of. By JAMES HADDON, M.A., Second Mathematical Master of King's College School. With Appendix, containing miscellaneous Investigations, and a Collection of Problems in various parts of Algebra. 2s.

86*. A KEY AND COMPANION to the above Book, forming an extensive repository of Solved Examples and Problems in Illustration of the various Expedients necessary in Algebraical Operations. Especially adapted for Self-Instruction. By J. R. YOUNG. 1s. 6d.

88. *EUCLID*, THE ELEMENTS OF : with many additional Propositions
89. and Explanatory Notes: to which is prefixed, an Introductory Essay on Logic. By HENRY LAW, C.E. 2s. 6d.‡

*** *Sold also separately, viz. :—*

88. EUCLID, The First Three Books. By HENRY LAW, C.E. 1s. 6d.
89. EUCLID, Books 4, 5, 6, 11, 12. By HENRY LAW, C.E. 1s. 6d.

☞ *The ‡ indicates that these vols. may be had strongly bound at 6d. extra.*

LONDON : CROSBY LOCKWOOD AND CO.,

Arithmetic, Geometry, Mathematics, etc., *continued.*

90. *ANALYTICAL GEOMETRY AND CONIC SECTIONS,* a Rudimentary Treatise on. By JAMES HANN, late Mathematical Master of King's College School, London. A New Edition, re-written and enlarged by J. R. YOUNG, formerly Professor of Mathematics at Belfast College. 2s.‡

91. *PLANE TRIGONOMETRY,* the Elements of. By JAMES HANN, formerly Mathematical Master of King's College, London. 1s. 6d.

92. *SPHERICAL TRIGONOMETRY,* the Elements of. By JAMES HANN. Revised by CHARLES H. DOWLING, C.E. 1s.
   ° Or with "The Elements of Plane Trigonometry," in One Volume, 2s. 6d. '

93. *MENSURATION AND MEASURING,* for Students and Practical Use. With the Mensuration and Levelling of Land for the Purposes of Modern Engineering. By T. BAKER, C.E. New Edition, with Corrections and Additions by E. NUGENT, C.E. Illustrated. 1s. 6d.

101*. *MEASURES, WEIGHTS, AND MONEYS OF ALL NA-TIONS,* and an Analysis of the Christian, Hebrew, and Mahometan Calendars. Entirely New Edition, revised and considerably enlarged. By W. S. B. WOOLHOUSE, F.R.A.S. [*In the Press.*

102. *INTEGRAL CALCULUS,* Rudimentary Treatise on the. By HOMERSHAM COX, B.A. Illustrated. 1s.

103. *INTEGRAL CALCULUS,* Examples on the. By JAMES HANN, late of King's College, London. Illustrated. 1s.

101. *DIFFERENTIAL CALCULUS,* Elements of the. By W. S. B. WOOLHOUSE, F.R.A.S., &c. 1s. 6d.

105. *MNEMONICAL LESSONS.* — GEOMETRY, ALGEBRA, AND TRIGONOMETRY, in Easy Mnemonical Lessons. By the Rev. THOMAS PENYNGTON KIRKMAN, M.A. 1s. 6d.

136. *ARITHMETIC,* Rudimentary, for the Use of Schools and Self-Instruction. By JAMES HADDON, M.A. Revised by ABRAHAM ARMAN. 1s. 6d.

137. A KEY TO HADDON'S RUDIMENTARY ARITHMETIC. By A. ARMAN. 1s. 6d.

168. *DRAWING AND MEASURING INSTRUMENTS.* Including—I. Instruments employed in Geometrical and Mechanical Drawing, and in the Construction, Copying, and Measurement of Maps and Plans. II. Instruments used for the purposes of Accurate Measurement, and for Arithmetical Computations. By J. F. HEATHER, M.A., late of the Royal Military Academy, Woolwich, Author of "Descriptive Geometry," &c., &c. Illustrated. 1s. 6d.

169. *OPTICAL INSTRUMENTS.* Including (more especially) Telescopes, Microscopes, and Apparatus for producing copies of Maps and Plans by Photography. By J. F. HEATHER, M.A. Illustrated. 1s. 6d.

170. *SURVEYING AND ASTRONOMICAL INSTRUMENTS.* Including—I. Instruments Used for Determining the Geometrical Features of a portion of Ground. II. Instruments Employed in Astronomical Observations. By J. F. HEATHER, M.A. Illustrated. 1s. 6d.
   ° *The above three volumes form an enlargement of the Author's original work,* "*Mathematical Instruments: their Construction, Adjustment, Testing, and Use,*" *the Thirteenth Edition of which is on sale, price 1s. 6d. (See No. 32 in the Series.)*

168.⎫
169.⎬ *MATHEMATICAL INSTRUMENTS.* By J. F. HEATHER, M.A. Enlarged Edition, for the most part entirely re-written. The 3 Parts as
170.⎭ above, in One thick Volume. With numerous Illustrations. 4s. 6d.‡

158. *THE SLIDE RULE, AND HOW TO USE IT;* containing full, easy, and simple Instructions to perform all Business Calculations with unexampled rapidity and accuracy. By CHARLES HOARE, C.E. With a Slide Rule in tuck of cover. 2s. 6d.‡

185. *THE COMPLETE MEASURER;* setting forth the Measurement of Boards, Glass, &c., &c.; Unequal-sided, Square-sided, Octagonal-sided, Round Timber and Stone, and Standing Timber. With a Table showing the solidity of hewn or eight-sided timber, or of any octagonal-sided column. Compiled for Timber-growers, Merchants, and Surveyors, Stonemasons, Architects, and others. By RICHARD HORTON. Third Edition, with valuable additions. 4s.; strongly bound in leather, 5s.

☞ *The ‡ indicates that these vols . may be had strongly bound at 6d. extra.*

Arithmetic, Geometry, Mathematics, etc., *continued.*

196. *THEORY OF COMPOUND INTEREST AND ANNUI-*
TIES; with Tables of Logarithms for the more Difficult Computations of
Interest, Discount, Annuities, &c. By FÉDOR THOMAN, of the Société Crédit
Mobilier, Paris. 4s.‡

199. *INTUITIVE CALCULATIONS;* or, Easy and Compendious
Methods of Performing the various Arithmetical Operations required in
Commercial and Business Transactions; together with Full Explanations of
Decimals and Duodecimals, several Useful Tables, and an Examination and
Discussion of the best Schemes for a Decimal Coinage. By DANIEL
O'GORMAN. Twenty-fifth Edition, corrected and enlarged by J. R. YOUNG,
formerly Professor of Mathematics in Belfast College. . 3s.‡

204. *MATHEMATICAL TABLES,* for Trigonometrical, Astronomical,
and Nautical Calculations; to which is prefixed a Treatise on Logarithms.
By HENRY LAW, C.E. Together with a Series of Tables for Navigation
and Nautical Astronomy. By J. R. YOUNG, formerly Professor of Mathematics in Belfast College. New Edition. 3s. 6d.‡

MISCELLANEOUS VOLUMES.

36. *A DICTIONARY OF TERMS used in ARCHITECTURE,*
BUILDING, ENGINEERING, MINING, METALLURGY, ARCHÆ-
OLOGY, the FINE ARTS, &c. By JOHN WEALE. Fifth Edition. Revised
by ROBERT HUNT, F.R.S., Keeper of Mining Records. Numerous Illustrations. 5s. cloth limp ; 6s. cloth boards.

50. *THE LAW OF CONTRACTS FOR WORKS AND SER-*
VICES. By DAVID GIBBONS. Third Edition, enlarged. 3s.‡

112. *MANUAL OF DOMESTIC MEDICINE.* By R. GOODING,
B.A., M.D. Intended as a Family Guide in all Cases of Accident and
Emergency. 2s.‡

112*. *MANAGEMENT OF HEALTH.* A Manual of Home and
Personal Hygiene. By the Rev. JAMES BAIRD, B.A. 1s.

150. *LOGIC,* Pure and Applied. By S. H. EMMENS. 1s. 6d.

152. *PRACTICAL HINTS FOR INVESTING MONEY.* With
an Explanation of the Mode of Transacting Business on the Stock Exchange.
By FRANCIS PLAYFORD, Sworn Broker. 1s. 6d.

153. *SELECTIONS FROM LOCKE'S ESSAYS ON THE*
HUMAN UNDERSTANDING. With Notes by S. H. EMMENS. 2s.

154. *GENERAL HINTS TO EMIGRANTS.* Containing Notices
of the various Fields for Emigration. With Hints on Preparation for
Emigrating, Outfits, &c., &c. With Directions and Recipes useful to the
Emigrant. With a Map of the World. 2s.

157. *THE EMIGRANT'S GUIDE TO NATAL.* By ROBERT
JAMES MANN, F.R.A.S., F.M.S. Second Edition, carefully corrected to
the present Date. Map. 2s.

193. *HANDBOOK OF FIELD FORTIFICATION,* intended for the
Guidance of Officers Preparing for Promotion, and especially adapted to the
requirements of Beginners. By Major W. W. KNOLLYS, F.R.G.S., 93rd
Sutherland Highlanders, &c. With 163 Woodcuts. 3s.‡

194. *THE HOUSE MANAGER :* Being a Guide to Housekeeping.
Practical Cookery, Pickling and Preserving, Household Work, Dairy
Management, the Table and Dessert, Cellarage of Wines, Home-brewing
and Wine-making, the Boudoir and Dressing-room, Travelling, Stable
Economy, Gardening Operations, &c. By AN OLD HOUSEKEEPER. 3s. 6d.‡

194. *HOUSE BOOK (The).* Comprising :—I. THE HOUSE MANAGER.
112. By an OLD HOUSEKEEPER. II. DOMESTIC MEDICINE. By RALPH GOODING,
& M.D. III. MANAGEMENT OF HEALTH. By JAMES BAIRD. In One Vol.,
112*. strongly half-bound. 6s.

☞ *The ‡ indicates that these vols. may be had strongly bound at 6d. extra.*

LONDON : CROSBY LOCKWOOD AND CO.,

EDUCATIONAL AND CLASSICAL SERIES.

HISTORY.

1. **England, Outlines of the History of;** more especially with reference to the Origin and Progress of the English Constitution. By WILLIAM DOUGLAS HAMILTON, F.S.A., of Her Majesty's Public Record Office. 4th Edition, revised. 5s.; cloth boards, 6s.

5. **Greece, Outlines of the History of;** in connection with the Rise of the Arts and Civilization in Europe. By W. DOUGLAS HAMILTON, of University College, London, and EDWARD LEVIEN, M.A., of Balliol College, Oxford. 2s. 6d.; cloth boards, 3s. 6d.

7. **Rome, Outlines of the History of:** from the Earliest Period to the Christian Era and the Commencement of the Decline of the Empire. By EDWARD LEVIEN, of Balliol College, Oxford. Map, 2s. 6d.; cl. bds. 3s. 6d.

9. **Chronology of History, Art, Literature, and Progress,** from the Creation of the World to the Conclusion of the Franco-German War. The Continuation by W. D. HAMILTON, F.S.A. 3s.; cloth boards, 3s. 6d.

50. **Dates and Events in English History,** for the use of Candidates in Public and Private Examinations. By the Rev. E. RAND. 1s.

ENGLISH LANGUAGE AND MISCELLANEOUS.

11. **Grammar of the English Tongue, Spoken and Written.** With an Introduction to the Study of Comparative Philology. By HYDE CLARKE, D.C.L. Fourth Edition. 1s. 6d.

11*. **Philology:** Handbook of the Comparative Philology of English, Anglo-Saxon, Frisian, Flemish or Dutch, Low or Platt Dutch, High Dutch or German, Danish, Swedish, Icelandic, Latin, Italian, French, Spanish, and Portuguese Tongues. By HYDE CLARKE, D.C.L. 1s.

12. **Dictionary of the English Language,** as Spoken and Written. Containing above 100,000 Words. By HYDE CLARKE, D.C.L. 3s. 6d.; cloth boards, 4s. 6d.; complete with the GRAMMAR, cloth bds., 5s. 6d.

48. **Composition and Punctuation,** familiarly Explained for those who have neglected the Study of Grammar. By JUSTIN BRENAN. 17th Edition. 1s. 6d.

49. **Derivative Spelling-Book:** Giving the Origin of Every Word from the Greek, Latin, Saxon, German, Teutonic, Dutch, French, Spanish, and other Languages; with their present Acceptation and Pronunciation. By J. ROWBOTHAM, F.R.A.S. Improved Edition. 1s. 6d.

51. **The Art of Extempore Speaking:** Hints for the Pulpit, the Senate, and the Bar. By M. BAUTAIN, Vicar-General and Professor at the Sorbonne. Translated from the French. 7th Edition, carefully corrected. 2s. 6d.

52. **Mining and Quarrying,** with the Sciences connected therewith. First Book of, for Schools. By J. H. COLLINS, F.G.S., Lecturer to the Miners' Association of Cornwall and Devon. 1s.

53. **Places and Facts in Political and Physical Geography,** for Candidates in Examinations. By the Rev. EDGAR RAND, B.A. 1s.

54. **Analytical Chemistry,** Qualitative and Quantitative, a Course of. To which is prefixed, a Brief Treatise upon Modern Chemical Nomenclature and Notation. By WM. W. PINK and GEORGE E. WEBSTER. 2s.

THE SCHOOL MANAGERS' SERIES OF READING BOOKS,

Adapted to the Requirements of the New Code. Edited by the Rev. A. R. GRANT, Rector of Hitcham, and Honorary Canon of Ely; formerly H.M. Inspector of Schools.

INTRODUCTORY PRIMER, 3d.

	s.	d.				s.	d.
FIRST STANDARD	0	6	FOURTH STANDARD	.	.	1	2
SECOND ,,	0	10	FIFTH ,,	.	.	1	6
THIRD ,,	1	0	SIXTH ,,	.	.	1	6

LESSONS FROM THE BIBLE. Part I. Old Testament. 1s.
LESSONS FROM THE BIBLE. Part II. New Testament, to which is added THE GEOGRAPHY OF THE BIBLE, for very young Children. By Rev. C. THORNTON FORSTER. 1s. 2d. *₊* Or the Two Parts in One Volume. 2s.

FRENCH.

24. **French Grammar.** With Complete and Concise Rules on the Genders of French Nouns. By G. L. STRAUSS, Ph.D. 1s. 6d.
25. **French-English Dictionary.** Comprising a large number of New Terms used in Engineering, Mining, &c. By ALFRED ELWES. 1s. 6d.
26. **English-French Dictionary.** By ALFRED ELWES. 2s.
25,26. **French Dictionary** (as above). Complete, in One Vol., 3s.; cloth boards, 3s. 6d. *₊* Or with the GRAMMAR, cloth boards, 4s. 6d.
47. **French and English Phrase Book :** containing Intro-ductory Lessons, with Translations, several Vocabularies of Words, a Collection of suitable Phrases, and Easy Familiar Dialogues. 1s. 6d.

GERMAN.

39. **German Grammar.** Adapted for English Students, from Heyse's Theoretical and Practical Grammar, by Dr. G. L. STRAUSS. 1s.
40. **German Reader :** A Series of Extracts, carefully culled from the most approved Authors of Germany; with Notes, Philological and Explanatory. By G. L. STRAUSS, Ph.D. 1s.
41. **German Triglot Dictionary.** By NICHOLAS ESTERHAZY S. A. HAMILTON. Part I. English-German-French. 1s.
42. **German Triglot Dictionary.** Part II. German-French-English. 1s.
43. **German Triglot Dictionary.** Part III. French-German-English. 1s.
41-43. **German Triglot Dictionary** (as above), in One Vol., 3s.; cloth boards, 4s. *₊* Or with the GERMAN GRAMMAR, cloth boards, 5s.

ITALIAN.

27. **Italian Grammar,** arranged in Twenty Lessons, with a Course of Exercises. By ALFRED ELWES. 1s. 6d.
28. **Italian Triglot Dictionary,** wherein the Genders of all the Italian and French Nouns are carefully noted down. By ALFRED ELWES. Vol. 1. Italian-English-French. 2s. 6d.
30. **Italian Triglot Dictionary.** By A. ELWES. Vol. 2. English-French-Italian. 2s. 6d.
32. **Italian Triglot Dictionary.** By ALFRED ELWES. Vol. 3. French-Italian-English. 2s. 6d.
28,30, **Italian Triglot Dictionary** (as above). In One Vol., 7s. 6d.
32.   Cloth boards.

SPANISH AND PORTUGUESE.

34. **Spanish Grammar,** in a Simple and Practical Form. With a Course of Exercises. By ALFRED ELWES. 1s. 6d.
35. **Spanish-English and English-Spanish Dictionary.** Including a large number of Technical Terms used in Mining, Engineering, &c., with the proper Accents and the Gender of every Noun. By ALFRED ELWES. 4s.; cloth boards, 5s. *₊* Or with the GRAMMAR, cloth boards, 6s.
55. **Portuguese Grammar,** in a Simple and Practical Form. With a Course of Exercises. By ALFRED ELWES. 1s. 6d.
56. **Portuguese-English and English-Portuguese Dictionary;** with the Genders of each Noun. By ALFRED ELWES.
[In preparation.

HEBREW.

46*. **Hebrew Grammar.** By Dr. BRESSLAU. 1s. 6d.
44. **Hebrew and English Dictionary,** Biblical and Rabbinical; containing the Hebrew and Chaldee Roots of the Old Testament Post-Rabbinical Writings. By Dr. BRESSLAU. 6s. *₊* Or with the GRAMMAR, 7s.
46. **English and Hebrew Dictionary.** By Dr. BRESSLAU. 3s.
44,46. **Hebrew Dictionary** (as above), in Two Vols., complete, with
46*.   the GRAMMAR, cloth boards, 12s.

LATIN.

19. **Latin Grammar.** Containing the Inflections and Elementary Principles of Translation and Construction. By the Rev. THOMAS GOODWIN, M.A., Head Master of the Greenwich Proprietary School. 1s.

20. **Latin-English Dictionary.** By the Rev. THOMAS GOODWIN, M.A. 2s.

22. **English-Latin Dictionary;** together with an Appendix of French and Italian Words which have their origin from the Latin. By the Rev. THOMAS GOODWIN, M.A. 1s. 6d.

20,22. **Latin Dictionary** (as above). Complete in One Vol., 3s. 6d.; cloth boards, 4s. 6d. *⋅* Or with the GRAMMAR, cloth boards, 5s. 6d.

LATIN CLASSICS. With Explanatory Notes in English.

1. **Latin Delectus.** Containing Extracts from Classical Authors, with Genealogical Vocabularies and Explanatory Notes, by H. YOUNG. 1s.

2. **Cæsaris Commentarii de Bello Gallico.** Notes, and a Geographical Register for the Use of Schools, by H. YOUNG. 2s.

3. **Cornelius Nepos.** With Notes. By H. YOUNG. 1s.

4. **Virgilii Maronis Bucolica et Georgica.** With Notes on the Bucolics by W. RUSHTON, M.A., and on the Georgics by H. YOUNG. 1s. 6d.

5. **Virgilii Maronis Æneis.** With Notes, Critical and Explanatory, by H. YOUNG. New Edition, revised and improved. With copious Additional Notes by Rev. T. H. L. LEARY, D.C.L., formerly Scholar of Brasenose College, Oxford. 3s.

5*. ———— Part 1. Books i.—vi., 1s. 6d.

5**. ———— Part 2. Books vii.—xii., 2s.

6. **Horace;** Odes, Epode, and Carmen Sæculare. Notes by H. YOUNG. 1s. 6d.

7. **Horace;** Satires, Epistles, and Ars Poetica. Notes by W. BROWNRIGG SMITH, M.A., F.R.G.S. 1s. 6d.

8. **Sallustii Crispi Catalina et Bellum Jugurthinum.** Notes, Critical and Explanatory, by W. M. DONNE, B.A., Trin. Coll., Cam. 1s. 6d.

9. **Terentii Andria et Heautontimorumenos.** With Notes, Critical and Explanatory, by the Rev. JAMES DAVIES, M.A. 1s. 6d.

10. **Terentii Adelphi, Hecyra, Phormio.** Edited, with Notes, Critical and Explanatory, by the Rev. JAMES DAVIES, M.A. 2s.

11. **Terentii Eunuchus, Comœdia.** Notes, by Rev. J. DAVIES, M.A. 1s. 6d.

12. **Ciceronis Oratio pro Sexto Roscio Amerino.** Edited, with an Introduction, Analysis, and Notes, Explanatory and Critical, by the Rev. JAMES DAVIES, M.A. 1s.

13. **Ciceronis Orationes in Catilinam, Verrem, et pro Archia.** With Introduction, Analysis, and Notes, Explanatory and Critical, by Rev. T. H. L. LEARY, D.C.L. formerly Scholar of Brasenose College, Oxford. 1s. 6d.

14. **Ciceronis Cato Major, Lælius, Brutus, sive de Senectute, de Amicitia, de Claris Oratoribus Dialogi.** With Notes by W. BROWNRIGG SMITH, M.A., F.R.G.S. 2s.

16. **Livy:** History of Rome. Notes by H. YOUNG and W. B. SMITH, M.A. Part 1. Books i., ii., 1s. 6d.

16*. ———— Part 2. Books iii., iv., v., 1s. 6d.

17. ———— Part 3. Books xxi., xxii., 1s. 6d.

19. **Latin Verse Selections,** from Catullus, Tibullus, Propertius, and Ovid. Notes by W. B. DONNE, M.A., Trinity College, Cambridge. 2s.

20. **Latin Prose Selections,** from Varro, Columella, Vitruvius, Seneca, Quintilian, Florus, Velleius Paterculus, Valerius Maximus Suetonius, Apuleius, &c. Notes by W. DONNE, M.A. 2s.

21. **Juvenalis Satiræ.** With Prolegomena and Notes by T. H. S. ESCOTT, B.A., Lecturer on Logic at King's College, London. 2s.

GREEK.

14. **Greek Grammar,** in accordance with the Principles and Philological Researches of the most eminent Scholars of our own day. By HANS CLAUDE HAMILTON. 1s. 6d.

15,17. **Greek Lexicon.** Containing all the Words in General Use, with their Significations, Inflections, and Doubtful Quantities. By HENRY R. HAMILTON. Vol. 1. Greek-English, 2s. 6d.; Vol. 2. English-Greek, 2s. Or the Two Vols. in One, 4s. 6d.: cloth boards, 5s.

14,15. **Greek Lexicon** (as above). Complete, with the GRAMMAR, in
17. One Vol., cloth boards, 6s.

GREEK CLASSICS. With Explanatory Notes in English.

1. **Greek Delectus.** Containing Extracts from Classical Authors, with Genealogical Vocabularies and Explanatory Notes, by H. YOUNG. New Edition, with an improved and enlarged Supplementary Vocabulary, by JOHN HUTCHISON, M.A., of the High School, Glasgow. 1s. 6d.

2, 3. **Xenophon's Anabasis;** or, The Retreat of the Ten Thousand. Notes and a Geographical Register, by H. YOUNG. Part 1. Books i. to iii., 1s. Part 2. Books iv. to vii., 1s.

4. **Lucian's Select Dialogues.** The Text carefully revised, with Grammatical and Explanatory Notes, by H. YOUNG. 1s. 6d.

5-12. **Homer, The Works of.** According to the Text of BAEUMLEIN. With Notes, Critical and Explanatory, drawn from the best and latest Authorities, with Preliminary Observations and Appendices, by T. H. L. LEARY, M.A., D.C.L.

THE ILIAD:	Part 1. Books i. to vi., 1s. 6d.	Part 3. Books xiii. to xviii., 1s. 6d.
	Part 2. Books vii. to xii., 1s. 6d.	Part 4. Books xix. to xxiv., 1s. 6d.
THE ODYSSEY:	Part 1. Books i. to vi., 1s. 6d	Part 3. Books xiii. to xviii., 1s. 6d.
	Part 2. Books vii. to xii., 1s. 6d.	Part 4. Books xix. to xxiv., and Hymns, 2s.

13. **Plato's Dialogues:** The Apology of Socrates, the Crito, and the Phædo. From the Text of C. F. HERMANN. Edited with Notes, Critical and Explanatory, by the Rev. JAMES DAVIES, M.A. 2s.

14-17. **Herodotus, The History of,** chiefly after the Text of GAISFORD. With Preliminary Observations and Appendices, and Notes, Critical and Explanatory, by T. H. L. LEARY, M.A., D.C.L.
      Part 1. Books i., ii. (The Clio and Euterpe), 2s.
      Part 2. Books iii., iv. (The Thalia and Melpomene), 2s.
      Part 3. Books v.-vii. (The Terpsichore, Erato, and Polymnia), 2s.
      Part 4. Books viii., ix. (The Urania and Calliope) and Index, 1s. 6d.

18. **Sophocles:** Œdipus Tyrannus. Notes by H. YOUNG. 1s.

20. **Sophocles:** Antigone. From the Text of DINDORF. Notes, Critical and Explanatory, by the Rev. JOHN MILNER, B.A. 2s.

23. **Euripides:** Hecuba and Medea. Chiefly from the Text of DINDORF. With Notes, Critical and Explanatory, by W. BROWNRIGG SMITH, M.A., F.R.G.S. 1s. 6d.

26. **Euripides:** Alcestis. Chiefly from the Text of DINDORF. With Notes, Critical and Explanatory, by JOHN MILNER, B.A. 1s. 6d.

30. **Æschylus:** Prometheus Vinctus: The Prometheus Bound. From the Text of DINDORF. Edited, with English Notes, Critical and Explanatory, by the Rev. JAMES DAVIES, M.A. 1s.

32. **Æschylus:** Septem Contra Thebes: The Seven against Thebes. From the Text of DINDORF. Edited, with English Notes, Critical and Explanatory, by the Rev. JAMES DAVIES, M.A. 1s.

40. **Aristophanes:** Acharnians. Chiefly from the Text of C. H. WEISE. With Notes, by C. S. T. TOWNSHEND, M.A. 1s. 6d.

41. **Thucydides:** History of the Peloponnesian War. Notes by H. YOUNG. Book 1. 1s.

42. **Xenophon's Panegyric on Agesilaus.** Notes and Introduction by LL. F. W. JEWITT. 1s. 6d.

43. **Demosthenes.** The Oration on the Crown and the Philippics. With English Notes. By Rev. T. H. L. LEARY, D.C.L., formerly Scholar of Brasenose College, Oxford. 1s. 6d.

LONDON, *December*, 1880.

A Catalogue of Books

INCLUDING MANY NEW AND STANDARD WORKS IN

ENGINEERING, ARCHITECTURE, AGRICULTURE,

MATHEMATICS, MECHANICS, SCIENCE, ETC.

PUBLISHED BY

CROSBY LOCKWOOD & CO.,

7, STATIONERS'-HALL COURT, LUDGATE HILL, E.C.

ENGINEERING, SURVEYING, ETC.

Humber's New Work on Water-Supply.

A COMPREHENSIVE TREATISE on the WATER-SUPPLY of CITIES and TOWNS. By WILLIAM HUMBER, A-M. Inst. C.E., and M. Inst. M.E. Illustrated with 50 Double Plates, 1 Single Plate, Coloured Frontispiece, and upwards of 250 Woodcuts, and containing 400 pages of Text, Imp. 4to, 6*l*. 6*s*. elegantly and substantially half-bound in morocco.

List of Contents :—

I. Historical Sketch of some of the means that have been adopted for the Supply of Water to Cities and Towns.—II. Water and the Foreign Matter usually associated with it.—III. Rainfall and Evaporation.—IV. Springs and the water-bearing formations of various districts.—V. Measurement and Estimation of the Flow of Water.—VI. On the Selection of the Source of Supply.—VII. Wells.—VIII Reservoirs.—IX. The Purification of Water.—X. Pumps.—XI. Pumping Machinery.—XII. Conduits.—XIII. Distribution of Water.—XIV. Meters, Service Pipes, and House Fittings.—XV. The Law and Economy of Water Works.—XVI. Constant and Intermittent Supply. —XVII. Description of Plates.—Appendices, giving Tables of Rates of Supply, Velocities, &c. &c., together with Specifications of several Works illustrated, among which will be found :—Aberdeen, Bideford, Canterbury, Dundee, Halifax, Lambeth, Rotherham, Dublin, and others.

" The most systematic and valuable work upon water supply hitherto produced in English, or in any other language . . . . Mr. Humber's work is characterised almost throughout by an exhaustiveness much more distinctive of French and German than of English technical treatises."—*Engineer.*

Humber's Great Work on Bridge Construction.

A COMPLETE and PRACTICAL TREATISE on CAST and WROUGHT-IRON BRIDGE CONSTRUCTION, including Iron Foundations. In Three Parts—Theoretical, Practical, and Descriptive. By WILLIAM HUMBER, A-M. Inst. C.E., and M. Inst. M.E. Third Edition, with 115 Double Plates. In 2 vols. imp. 4to, 6*l*. 16*s*. 6*d*. half-bound in morocco.

" A book—and particularly a large and costly treatise like Mr. Humber's—which has reached its third edition may certainly be said to have established its own reputation."—*Engineering.*

B

Humber's Modern Engineering.

A RECORD of the PROGRESS of MODERN ENGINEER-ING. First Series. Comprising Civil, Mechanical, Marine, Hydraulic, Railway, Bridge, and other Engineering Works, &c. By WILLIAM HUMBER, A-M. Inst. C.E., &c. Imp. 4to, with 36 Double Plates, drawn to a large scale, and Portrait of John Hawkshaw C.E., F.R.S., &c., and descriptive Letter-press, Specifications, &c. 3*l.* 3*s.* half morocco.

List of the Plates and Diagrams.

Victoria Station and Roof, L. B. & S. C. R. (8 plates); Southport Pier (2 plates); Victoria Station and Roof, L. C. & D. and G. W. R. (6 plates); Roof of Cremorne Music Hall; Bridge over G. N. Railway; Roof of Station, Dutch Rhenish Rail (2 plates); Bridge over the Thames, West London Extension Railway (5 plates); Armour Plates; Suspension Bridge, Thames (4 plates): The Allen Engine; Suspension Bridge, Avon (3 plates); Underground Railway (3 plates).

" Handsomely lithographed and printed. It will find favour with many who desire to preserve in a permanent form copies of the plans and specifications prepared for the guidance of the contractors for many important engineering works."—*Engineer.*

HUMBER'S RECORD OF MODERN ENGINEERING. Second Series. Imp. 4to, with 36 Double Plates, Portrait of Robert Stephenson, C.E., &c., and descriptive Letterpress, Specifications, &c. 3*l.* 3*s.* half morocco.

List of the Plates and Diagrams.

Birkenhead Docks, Low Water Basin (15 plates); Charing Cross Station Roof, C. C. Railway (3 plates); Digswell Viaduct, G. N. Railway; Robbery Wood Viaduct, G. N. Railway; Iron Permanent Way; Clydach Viaduct, Merthyr, Tredegar, and Abergavenny Railway; Ebbw Viaduct, Merthyr, Tredegar, and Abergavenny Railway; College Wood Viaduct, Cornwall Railway; Dublin Winter Palace Roof (3 plates); Bridge over the Thames, L. C. and D. Railway (6 plates); Albert Harbour, Greenock (4 plates).

HUMBER'S RECORD OF MODERN ENGINEERING. Third Series. Imp. 4to, with 40 Double Plates, Portrait of J. R. M'Clean, Esq., late Pres. Inst. C.E, and descriptive Letterpress, Specifications, &c. 3*l.* 3*s.* half morocco.

List of the Plates and Diagrams.

MAIN DRAINAGE, METROPOLIS. — *North Side.*—Map showing Interception of Sewers; Middle Level Sewer (2 plates); Outfall Sewer, Bridge over River Lea (3 plates); Outfall Sewer, Bridge over Marsh Lane, North Woolwich Railway, and Bow and Barking Railway Junction; Outfall Sewer, Bridge over Bow and Barking Railway (3 plates); Outfall Sewer, Bridge over East London Waterworks' Feeder (2 plates); Outfall Sewer, Reservoir (2 plates); Outfall Sewer, Tumbling Bay and Outlet; Outfall Sewer, Penstocks. *South Side.*—Outfall Sewer, Bermondsey Branch (2 plates); Outfall Sewer, Reservoir and Outlet (4 plates); Outfall Sewer, Filth Hoist; Sections of Sewers (North and South Sides).

THAMES EMBANKMENT.— Section of River Wall; Steamboat Pier, Westminster (2 plates); Landing Stairs between Charing Cross and Waterloo Bridges; York Gate (2 plates); Overflow and Outlet at Savoy Street Sewer (3 plates); Steamboat Pier, Waterloo Bridge (3 plates); Junction of Sewers, Plans and Sections; Gullies, Plans and Sections; Rolling Stock; Granite and Iron Forts.

HUMBER'S RECORD OF MODERN ENGINEERING. Fourth Series. Imp. 4to, with 36 Double Plates, Portrait of John Fowler, Esq., late Pres. Inst. C.E., and descriptive Letterpress, Specifications, &c. 3*l.* 3*s.* half morocco.

List of the Plates and Diagrams.

Abbey Mills Pumping Station, Main Drainage, Metropolis (4 plates); Barrow Docks (5 plates); Manquis Viaduct, Santiago and Valparaiso Railway (2 plates); Adam's Locomotive, St. Helen's Canal Railway (2 plates); Cannon Street Station Roof, Charing Cross Railway (3 plates); Road Bridge over the River Moka (2 plates); Telegraphic Apparatus for Mesopotamia; Viaduct over the River Wye, Midland Railway (3 plates); St. German's Viaduct, Cornwall Railway (2 plates); Wrought-Iron Cylinder for Diving Bell; Millwall Docks (6 plates); Milroy's Patent Excavator, Metropolitan District Railway (6 plates); Harbours, Ports, and Breakwaters (3 plates).

Strains, Formulæ & Diagrams for Calculation of.

A HANDY BOOK for the CALCULATION of STRAINS in GIRDERS and SIMILAR STRUCTURES, and their STRENGTH ; consisting of Formulæ and Corresponding Diagrams, with numerous Details for Practical Application, &c. By WILLIAM HUMBER, A-M. Inst. C. E., &c. Third Edition. With nearly 100 Woodcuts and 3 Plates, Crown 8vo, 7s. 6d. cloth.

"The arrangement of the matter in this little volume is as convenient as it well could be. . . . . The system of employing diagrams as a substitute for complex computations is one justly coming into great favour, and in that respect Mr. Humber's volume is fully up to the times."—*Engineering.*

"The formulæ are neatly expressed, and the diagrams good."—*Athenæum.*

Strains.

THE STRAINS ON STRUCTURES OF IRONWORK ; with Practical Remarks on Iron Construction. By F. W. SHEILDS, M. Inst. C.E. Second Edition, with 5 Plates. Royal 8vo, 5s. cloth.

"The student cannot find a better little book on this subject than that written by Mr. Sheilds."—*Engineer.*

Barlow on the Strength of Materials, enlarged.

A TREATISE ON THE STRENGTH OF MATERIALS, with Rules for application in Architecture, the Construction of Suspension Bridges, Railways, &c. ; and an Appendix on the Power of Locomotive Engines, and the effect of Inclined Planes and Gradients. By PETER BARLOW, F.R.S. A New Edition, revised by his Sons, P. W. BARLOW, F.R.S., and W. H. BARLOW, F.R.S. The whole arranged and edited by W. HUMBER, A-M. Inst. C.E. 8vo, 400 pp., with 19 large Plates, 18s. cloth.

"The best book on the subject which has yet appeared. . . . . We know of no work that so completely fulfils its mission."—*English Mechanic.*

"The standard treatise upon this particular subject."—*Engineer.*

Strength of Cast Iron, &c.

A PRACTICAL ESSAY on the STRENGTH of CAST IRON and OTHER METALS. By THOMAS TREDGOLD, C.E. Fifth Edition. To which are added, Experimental Researches on the Strength and other Properties of Cast Iron, by E. HODGKINSON, F.R.S. With 9 Engravings and numerous Woodcuts. 8vo, 12s. cloth. **** HODGKINSON'S RESEARCHES, separate, price 6s.

Hydraulics.

HYDRAULIC TABLES, CO-EFFICIENTS, and FORMULÆ for finding the Discharge of Water from Orifices, Notches, Weirs, Pipes, and Rivers. With New Formulæ, Tables, and General Information on Rain-fall, Catchment-Basins, Drainage, Sewerage, Water Supply for Towns and Mill Power. By JOHN NEVILLE, Civil Engineer, M.R.I.A. Third Edition, carefully revised, with considerable Additions. Numerous Illustrations. Cr. 8vo, 14s. cloth.

"Undoubtedly an exceedingly useful and elaborate compilation."—*Iron.*

"Alike valuable to students and engineers in practice."—*Mining Journal.*

River Engineering.

RIVER BARS : Notes on the Causes of their Formation, and on their Treatment by Induced Tidal Scour, with a Description of the Successful Reduction by this Method of the Bar at Dublin. By I. J. MANN, Assistant Engineer to the Dublin Port and Docks Board. With Illustrations. Demy 8vo. [*In the press.*

Levelling.

A TREATISE on the PRINCIPLES and PRACTICE of LEVELLING; showing its Application to Purposes of Railway and Civil Engineering, in the Construction of Roads; with Mr. TELFORD's Rules for the same. By FREDERICK W. SIMMS, F.G.S., M. Inst. C.E. Sixth Edition, very carefully revised, with the addition of Mr. LAW's Practical Examples for Setting out Railway Curves, and Mr. TRAUTWINE's Field Practice of Laying out Circular Curves. With 7 Plates and numerous Woodcuts. 8vo, 8s. 6d. cloth.   *₊* TRAUTWINE on Curves, separate, 5s.

"The text-book on levelling in most of our engineering schools and colleges."—*Engineer.*

Practical Tunnelling.

PRACTICAL TUNNELLING : Explaining in detail the Setting out of the Works, Shaft-sinking and Heading-Driving, Ranging the Lines and Levelling under Ground, Sub-Excavating, Timbering, and the Construction of the Brickwork of Tunnels with the amount of labour required for, and the Cost of, the various portions of the work. By F. W. SIMMS, M. Inst. C.E. Third Edition, Revised and Extended. By D. KINNEAR CLARK, M.I.C.E. Imp. 8vo, with 21 Folding Plates and numerous Wood Engravings, 30s. cloth.

"It has been regarded from the first as a text-book of the subject. . . . Mr. Clark has added immensely to the value of the book."—*Engineer.*

Steam.

STEAM AND THE STEAM ENGINE, Stationary and Portable. Being an Extension of Sewell's Treatise on Steam. By D. KINNEAR CLARK, M.I.C.E. Second Edition. 12mo, 4s. cloth.

Civil and Hydraulic Engineering.

CIVIL ENGINEERING (THE RUDIMENTS OF). By HENRY LAW, M. Inst. C.E. Including a Treatise on Hydraulic Engineering, by GEORGE R. BURNELL, M.I.C.E. Sixth Edition, Revised, with large additions on Recent Practice in Civil Engineering, by D. KINNEAR CLARK, M. Inst. C.E.   [*In the press.*

Gas-Lighting.

COMMON SENSE FOR GAS-USERS : a Catechism of Gas-Lighting for Householders, Gasfitters, Millowners, Architects, Engineers, &c. By R. WILSON, C.E. 2nd Edition. Cr. 8vo, 2s. 6d.

Bridge Construction in Masonry, Timber, & Iron.

EXAMPLES OF BRIDGE AND VIADUCT CONSTRUCTION OF MASONRY, TIMBER, AND IRON ; consisting of 46 Plates from the Contract Drawings or Admeasurement of select Works. By W. DAVIS HASKOLL, C.E. Second Edition, with the addition of 554 Estimates, and the Practice of Setting out Works, with 6 pages of Diagrams. Imp. 4to, 2l. 12s. 6d. half-morocco.

"A work of the present nature by a man of Mr. Haskoll's experience, must prove invaluable. The tables of estimates considerably enhance its value."—*Engineering.*

Earthwork.

EARTHWORK TABLES, showing the Contents in Cubic Yards of Embankments, Cuttings, &c., of Heights or Depths up to an average of 80 feet. By JOSEPH BROADBENT, C.E., and FRANCIS CAMPIN, C.E. Cr. 8vo, oblong, 5s. cloth.

Tramways and their Working.

TRAMWAYS : their CONSTRUCTION and WORKING. Containing a Comprehensive History of the System ; an exhaustive Analysis of the Various Modes of Traction, including Horse Power, Steam, Heated Water, and Compressed Air ; a Description of the varieties of Rolling Stock ; and ample Details of Cost and Working Expenses, with Special Reference to the Tramways of the United Kingdom. By D. KINNEAR CLARK, M. I. C. E., Author of ' Railway Machinery,' &c., in one vol. 8vo, with numerous Illustrations and thirteen folding Plates, 18s. cloth.

" All interested in tramways must refer to it, as all railway engineers have turned to the author's work ' Railway Machinery.' "—*The Engineer.*

" Mr. Clark's book is indispensable for the students of the subject."—*The Builder.*

Pioneer Engineering.

PIONEER ENGINEERING. A Treatise on the Engineering Operations connected with the Settlement of Waste Lands in New Countries. By EDWARD DOBSON, A.I.C.E. With Plates and Wood Engravings. Revised Edition. 12mo, 5s. cloth.

" A workmanlike production, and one without possession of which no man should start to encounter the duties of a pioneer engineer."—*Athenæum.*

" There is much in the book to render it very useful to an engineer proceeding to the colonies."—*Engineer.*

Steam Engine.

TEXT-BOOK ON THE STEAM ENGINE. By T. M. GOODEVE, M.A., Barrister-at-Law, Author of " The Principles of Mechanics," " The Elements of Mechanism," &c. Third Edition. With numerous Illustrations. Crown 8vo, 6s. cloth.

" Professor Goodeve has given us a treatise on the steam engine, which will bear comparison with anything written by Huxley or Maxwell, and we can award it no higher praise."—*Engineer.*

" Mr. Goodeve's text-book is a work of which every young engineer should possess himself."—*Mining Journal.*

Steam.

THE SAFE USE OF STEAM : containing Rules for Unprofessional Steam Users. By an ENGINEER. 4th Edition. Sewed, 6d.

" If steam-users would but learn this little book by heart, boiler explosions would become sensations by their rarity."—*English Mechanic.*

Works of Construction.

MATERIALS AND CONSTRUCTION : a Theoretical and Practical Treatise on the Strains, Designing, and Erection of Works of Construction. By FRANCIS CAMPIN, C.E., Author of " A Practical Treatise on Mechanical Engineering ; " " The Principles and Construction of Machinery," &c. With Numerous Illustrations. 12mo, 3s. 6d. cloth boards. [*Just published.*

Iron Bridges, Girders, Roofs, &c.

A TREATISE ON THE APPLICATION OF IRON TO THE CONSTRUCTION OF BRIDGES, GIRDERS, ROOFS, AND OTHER WORKS. By FRANCIS CAMPIN, C.E. Second Edition, Revised and Corrected. 12mo, 3s. cloth.

Construction of Iron Beams, Pillars, &c.

IRON AND HEAT ; exhibiting the Principles concerned in the construction of Iron Beams, Pillars, and Bridge Girders, and the Action of Heat in the Smelting Furnace. By J. ARMOUR, C.E. 3s.

Oblique Arches.

A PRACTICAL TREATISE ON THE CONSTRUCTION of OBLIQUE ARCHES. By JOHN HART. 3rd Ed. Imp. 8vo, 8s. cloth.

Oblique Bridges.

A PRACTICAL and THEORETICAL ESSAY on OBLIQUE BRIDGES, with 13 large Plates. By the late GEO. WATSON BUCK, M.I.C.E. Third Edition, revised by his Son, J. H. WATSON BUCK, M.I.C.E. ; and with the addition of Description to Diagrams for Facilitating the Construction of Oblique Bridges, by W. H. BARLOW, M.I.C.E. Royal 8vo, 12s. cloth. [*Just published.*
"The standard text book for all engineers regarding skew arches is Mr. Buck's treatise and it would be impossible to consult a better."—*Engineer.*

Gas and Gasworks.

THE CONSTRUCTION OF GASWORKS AND THE MANUFACTURE AND DISTRIBUTION OF COAL-GAS. Originally written by SAMUEL HUGHES, C.E. Sixth Edition. Re-written and much Enlarged, by WILLIAM RICHARDS, C.E. With 72 Woodcuts. 12mo, 5s. cloth boards. [*Just published.*

Waterworks for Cities and Towns.

WATERWORKS for the SUPPLY of CITIES and TOWNS, with a Description of the Principal Geological Formations of England as influencing Supplies of Water. By S. HUGHES. 4s. 6d. cloth.

Locomotive-Engine Driving.

LOCOMOTIVE-ENGINE DRIVING ; a Practical Manual for Engineers in charge of Locomotive Engines. By MICHAEL REYNOLDS, M.S.E., formerly Locomotive Inspector L. B. and S. C. R. Fourth Edition, greatly enlarged. Comprising A KEY TO THE LOCOMOTIVE ENGINE. With Illustrations and Portrait of Author. Crown 8vo, 4s. 6d. cloth.
"Mr. Reynolds deserves the title of the engine driver's friend."—*Railway News.*
"Mr. Reynolds has supplied a want, and has supplied it well. We can confidently recommend the book not only to the practical driver, but to every one who takes an interest in the performance of locomotive engines."—*Engineer.*

The Engineer, Fireman, and Engine-Boy.

THE MODEL LOCOMOTIVE ENGINEER, FIREMAN, AND ENGINE-BOY : comprising a Historical Notice of the Pioneer Locomotive Engines and their Inventors, with a project for the establishment of Certificates of Qualification in the Running Service of Railways. By MICHAEL REYNOLDS, Author of "Locomotive-Engine Driving." Crown 8vo, 4s. 6d. cloth.
"From the technical knowledge of the author it will appeal to the railway man of to-day more forcibly than anything written by Dr. Smiles."—*English Mechanic.*

Stationary Engine Driving.

STATIONARY ENGINE DRIVING. A Practical Manual fo Engineers in Charge of Stationary Engines. By MICHAEL REYNOLDS ("The Engine-Driver's Friend"), Author of "Locomotive Engine Driving," &c. With Plates and Woodcuts, and Steel Portrait of James Watt. Crown 8vo, 4s. 6d. cloth. [*Just published.*

Engine-Driving Life.

ENGINE-DRIVING LIFE ; or Stirring Adventures and Incidents in the Lives of Locomotive Engine-Drivers. By MICHAEL REYNOLDS. Crown 8vo, 2s. cloth. [*Just published.*

Fire Engineering.

FIRES, FIRE-ENGINES, AND FIRE BRIGADES. With a History of Fire-Engines, their Construction, Use, and Management; Remarks on Fire-Proof Buildings, and the Preservation of Life from Fire; Statistics of the Fire Appliances in English Towns; Foreign Fire Systems; Hints on Fire Brigades, &c., &c. By CHARLES F. T. YOUNG, C.E. With numerous Illustrations, handsomely printed, 544 pp., demy 8vo, 1l. 4s. cloth.

"We can most heartily commend this book."—*Engineering*.

"Mr. Young's book on 'Fire Engines and Fire Brigades' contains a mass of information, which has been collected from a variety of sources. The subject is so intensely interesting and useful that it demands consideration."—*Building News*.

Trigonometrical Surveying.

AN OUTLINE OF THE METHOD OF CONDUCTING A TRIGONOMETRICAL SURVEY, for the Formation of Geographical and Topographical Maps and Plans, Military Reconnaissance, Levelling, &c., with the most useful Problems in Geodesy and Practical Astronomy. By LIEUT.-GEN. FROME, R.E., late Inspector-General of Fortifications. Fourth Edition, Enlarged, and partly Re-written. By CAPTAIN CHARLES WARREN, R.E. With 19 Plates and 115 Woodcuts, royal 8vo, 16s. cloth.

Tables of Curves.

TABLES OF TANGENTIAL ANGLES and MULTIPLES for setting out Curves from 5 to 200 Radius. By ALEXANDER BEAZELEY, M. Inst. C.E. Second Edition. Printed on 48 Cards, and sold in a cloth box, waistcoat-pocket size, 3s. 6d.

"Each table is printed on a small card, which, being placed on the theodolite, leaves the hands free to manipulate the instrument—no small advantage as regards the rapidity of work."—*Engineer*.

"Very handy; a man may know that all his day's work must fall on two of these cards, which he puts into his own card-case, and leaves the rest behind."—*Athenæum*.

Engineering Fieldwork.

THE PRACTICE OF ENGINEERING FIELDWORK, applied to Land and Hydraulic, Hydrographic, and Submarine Surveying and Levelling. Second Edition, revised, with considerable additions, and a Supplement on WATERWORKS, SEWERS, SEWAGE, and IRRIGATION. By W. DAVIS HASKOLL, C.E. Numerous folding Plates. In One Volume, demy 8vo, 1l. 5s., cloth boards.

Large Tunnel Shafts.

THE CONSTRUCTION OF LARGE TUNNEL SHAFTS. A Practical and Theoretical Essay. By J. H. WATSON BUCK, M. Inst. C.E., Resident Engineer, London and North-Western Railway. Illustrated with Folding Plates. Royal 8vo, 12s. cloth. [*Just published*.

"Many of the methods given are of extreme practical value to the mason, and the observations on the form of arch, the rules for ordering the stone, and the construction of the templates, will be found of considerable use. We commend the book to the engineering profession, and to all who have to build similar shafts."—*Building News*.

"Will be regarded by civil engineers as of the utmost value, and calculated to save much time and obviate many mistakes."—*Colliery Guardian*.

Survey Practice.

AID TO SURVEY PRACTICE: for Reference in Surveying, Levelling, Setting-out and in Route Surveys of Travellers by Land and Sea. With Tables, Illustrations, and Records. By LOWIS D'A. JACKSON, A-M.I.C.E. Author of "Hydraulic Manual and Statistics," "Canal and Culvert Tables," &c. Large crown, 8vo, 12s. 6d., cloth.   [*Just published.*

"Mr. Jackson has produced a valuable *vade-mecum* for the surveyor. We can recommend this book as containing an admirable supplement to the teaching of the accomplished surveyor."—*Athenæum.*

"A general text book was wanted, and we are able to speak with confidence of Mr. Jackson's treatise. , . . We cannot recommend to the student who knows something of the mathematical principles of the subject a better course than to fortify his practice in the field under a competent surveyor with a study of Mr. Jackson's useful manual. The field records illustrate every kind of survey, and will be found an essential aid to the student."—*Building News.*

"The author brings to his work a fortunate union of theory and practical experience which, aided by a clear and lucid style of writing, renders the book both a very useful one and very agreeable to read."—*Builder.*

Sanitary Work.

SANITARY WORK IN THE SMALLER TOWNS AND IN VILLAGES. Comprising:—1. Some of the more Common Forms of Nuisance and their Remedies; 2. Drainage; 3. Water Supply. By CHAS. SLAGG, Assoc. Inst. C.E. Crown 8vo, 3s. cloth.

"A very useful book, and may be safely recommended. The author has had practical experience in the works of which he treats."—*Builder.*

Locomotives.

LOCOMOTIVE ENGINES, A Rudimentary Treatise on. Comprising an Historical Sketch and Description of the Locomotive Engine. By G. D. DEMPSEY, C.E. With large additions treating of the MODERN LOCOMOTIVE, by D. KINNEAR CLARK, C.E., M.I.C.E., Author of "Tramways, their Construction and Working," &c., &c. With numerous Illustrations. 12mo. 3s. 6d. cloth boards.

"The student cannot fail to profit largely by adopting this as his preliminary textbook."—*Iron and Coal Trades Review.*

"Seems a model of what an elementary technical book should be."—*Academy.*

Fuels and their Economy.

FUEL, its Combustion and Economy; consisting of an Abridgment of "A Treatise on the Combustion of Coal and the Prevention of Smoke." By C. W. WILLIAMS, A.I.C.E. With extensive additions on Recent Practice in the Combustion and Economy of Fuel—Coal, Coke, Wood, Peat, Petroleum, &c.; by D. KINNEAR CLARK, C.E., M.I.C.E. Second Edition, revised. With numerous Illustrations. 12mo. 4s. cloth boards. [*Just published.*

"Students should buy the book and read it, as one of the most complete and satisfactory treatises on the combustion and economy of fuel to be had."—*Engineer.*

Roads and Streets.

THE CONSTRUCTION OF ROADS AND STREETS. In Two Parts. I. The Art of Constructing Common Roads. By HENRY LAW, C.E. Revised and Condensed by D. KINNEAR CLARK, C.E.—II. Recent Practice in the Construction of Roads and Streets: including Pavements of Stone, Wood, and Asphalte. By D. KINNEAR CLARK, C.E., M.I.C.E. Second Edition, revised. 12mo, 5s. cloth.

"A book which every borough surveyor and engineer must possess, and which will be of considerable service to architects, builders, and property owners generally."—*Building News.*

Sewing Machine (The).

SEWING MACHINERY; being a Practical Manual of the Sewing Machine, comprising its History and Details of its Construction, with full Technical Directions for the Adjusting of Sewing Machines. By J. W. URQUHART, Author of "Electro Plating: a Practical Manual;" "Electric Light: its Production and Use." With Numerous Illustrations. 12mo, 2s. 6d. cloth boards.   [*Just published.*

Field-Book for Engineers.

THE ENGINEER'S, MINING SURVEYOR'S, and CONTRACTOR'S FIELD-BOOK. By W. DAVIS HASKOLL, C.E. Consisting of a Series of Tables, with Rules, Explanations of Systems, and Use of Theodolite for Traverse Surveying and Plotting the Work with minute accuracy by means of Straight Edge and Set Square only; Levelling with the Theodolite, Casting out and Reducing Levels to Datum, and Plotting Sections in the ordinary manner; Setting out Curves with the Theodolite by Tangential Angles and Multiples with Right and Left-hand Readings of the Instrument; Setting out Curves without Theodolite on the System of Tangential Angles by Sets of Tangents and Offsets; and Earthwork Tables to 80 feet deep, calculated for every 6 inches in depth. With numerous woodcuts. 4th Edition, enlarged. Cr. 8vo. 12s. cloth.

"The book is very handy, and the author might have added that the separate tables of sines and tangents to every minute will make it useful for many other purposes, the genuine traverse tables existing all the same."—*Athenæum.*

"Cannot fail, from its portability and utility, to be extensively patronised by the engineering profession."—*Mining Journal.*

Earthwork, Measurement and Calculation of.

A MANUAL on EARTHWORK. By ALEX. J. S. GRAHAM, C.E., Resident Engineer, Forest of Dean Central Railway. With numerous Diagrams. 18mo, 2s. 6d. cloth.

"As a really handy book for reference, we know of no work equal to it; and the railway engineers and others employed in the measurement and calculation of earthwork will find a great amount of practical information very admirably arranged, and available for general or rough estimates, as well as for the more exact calculations required in the engineers' contractor's offices."—*Artizan.*

Drawing for Engineers, &c.

THE WORKMAN'S MANUAL OF ENGINEERING DRAWING. By JOHN MAXTON, Instructor in Engineering Drawing, Royal Naval College, Greenwich, formerly of R. S. N. A., South Kensington. Fourth Edition, carefully revised. With upwards of 300 Plates and Diagrams. 12mo, cloth, strongly bound, 4s.

"A copy of it should be kept for reference in every drawing office."—*Engineering.*

"Indispensable for teachers of engineering drawing."—*Mechanics' Magazine.*

Weale's Dictionary of Terms.

A DICTIONARY of TERMS used in ARCHITECTURE, BUILDING, ENGINEERING, MINING, METALLURGY, ARCHÆOLOGY, the FINE ARTS, &c. By JOHN WEALE. Fifth Edition, revised by ROBERT HUNT, F.R.S., Keeper of Mining Records, Editor of "Ure's Dictionary of Arts." 12mo, 6s. cl. bds.

"The best small technological dictionary in the language."—*Architect.*

"The absolute accuracy of a work of this character can only be judged of after extensive consultation, and from our examination it appears very correct and very complete."—*Mining Journal.*

MINING, METALLURGY, ETC.

Metalliferous Minerals and Mining.

A TREATISE ON METALLIFEROUS MINERALS AND MINING. By D. C. DAVIES, F.G.S., author of "A Treatise on Slate and Slate Quarrying." With numerous wood engravings. Second Edition, revised. Cr. 8vo. 12s. 6d. cloth. [*Just published.*

"Without question, the most exhaustive and the most practically useful work we have seen ; the amount of information given is enormous, and it is given concisely and intelligibly."—*Mining Journal.*

"The volume is one which no student of mineralogy should be without."—*Colliery Guardian.*

"The author has gathered together from all available sources a vast amount of really useful information. As a history of the present state of mining throughout the world this book has a real value, and it supplies an actual want, for no such information has hitherto been brought together within such limited space."—*Athenæum.*

Slate and Slate Quarrying.

A TREATISE ON SLATE AND SLATE QUARRYING, Scientific, Practical, and Commercial. By D. C. DAVIES, F.G.S., Mining Engineer, &c. With numerous Illustrations and Folding Plates. Second Edition, carefully revised. 12mo, 3s. 6d. cloth boards.

"Mr. Davies has written a useful and practical hand-book on an important industry, with all the conditions and details of which he appears familiar."—*Engineering.*

"The work is illustrated by actual practice, and is unusually thorough and lucid. . . . Mr. Davies has completed his work with industry and skill."—*Builder.*

Metallurgy of Iron.

A TREATISE ON THE METALLURGY OF IRON : containing Outlines of the History of Iron Manufacture, Methods of Assay, and Analyses of Iron Ores, Processes of Manufacture of Iron and Steel, &c. By H. BAUERMAN, F.G.S., Associate of the Royal School of Mines. With numerous Illustrations. Fourth Edition, revised and much enlarged. 12mo, cloth boards, 5s.

"Has the merit of brevity and conciseness, as to less important points, while all material matters are very fully and thoroughly entered into."—*Standard.*

Manual of Mining Tools.

MINING TOOLS. For the use of Mine Managers, Agents, Mining Students, &c. By WILLIAM MORGANS, Lecturer on Practical Mining at the Bristol School of Mines. Volume of Text. 12mo, 3s. With an Atlas of Plates, containing 235 Illustrations. 4to, 6s. Together, 9s. cloth boards.

"Students in the Science of Mining, and Overmen, Captains, Managers, and Viewers may gain practical knowledge and useful hints by the study of Mr. Morgans' Manual."—*Colliery Guardian.*

Mining, Surveying and Valuing.

THE MINERAL SURVEYOR AND VALUER'S COMPLETE GUIDE, comprising a Treatise on Improved Mining Surveying, with new Traverse Tables ; and Descriptions of Improved Instruments ; also an Exposition of the Correct Principles of Laying out and Valuing Home and Foreign Iron and Coal Mineral Properties. By WILLIAM LINTERN, Mining and Civil Engineer. With four Plates of Diagrams, Plans, &c., 12mo, 4s. cloth.

"Contains much valuable information given in a small compass, and which, as far as we have tested it, is thoroughly trustworthy."—*Iron and Coal Trades Review.*

*** The above, bound with THOMAN'S TABLES. (See page 20.) Price 7s. 6d. cloth.

Coal and Coal Mining.

COAL AND COAL MINING: a Rudimentary Treatise on.   By
WARINGTON W. SMYTH, M.A., F.R.S., &c., Chief Inspector
of the Mines of the Crown.   Fifth edition, revised and corrected.
12mo, with numerous Illustations, 4s. cloth boards.

"Every portion of the volume appears to have been prepared with much care, and
as an outline is given of every known coal-field in this and other countries, as well as
of the two principal methods of working, the book will doubtless interest a very
large number of readers."—*Mining Journal.*

Underground Pumping Machinery.

MINE DRAINAGE ; being a Complete and Practical Treatise
on Direct-Acting Underground Steam Pumping Machinery, with
a Description of a large number of the best known Engines, their
General Utility and the Special Sphere of their Action, the Mode
of their Application, and their merits compared with other forms
of Pumping Machinery.   By STEPHEN MICHELL, Joint-Author of
" The Cornish System of Mine Drainage."   8vo.   [*Nearly ready.*

NAVAL ARCHITECTURE, NAVIGATION, ETC.

Pocket Book for Naval Architects & Shipbuilders.

THE NAVAL ARCHITECT'S AND SHIPBUILDER'S
POCKET BOOK OF FORMULÆ, RULES, AND TABLES
AND MARINE ENGINEER'S AND SURVEYOR'S HANDY
BOOK OF REFERENCE.   By CLEMENT MACKROW, Naval
Draughtsman, Associate of the Institution of Naval Architects.
With numerous Diagrams, &c.   Fcap., strongly bound in leather,
with elastic strap for pocket, 12s. 6d.

" Should be used by all who are engaged in the construction or design of vessels."
—*Engineer.*
" There is scarcely a subject on which a naval architect or shipbuilder can require
to refresh his memory which will not be found within the covers of Mr. Mackrow's
book."—*English Mechanic.*
" Mr. Mackrow has compressed an extraordinary amount of information into this
useful volume."—*Athenæum.*

Grantham's Iron Ship-Building.

ON IRON SHIP-BUILDING ; with Practical Examples and
Details.   Fifth Edition.   Imp. 4to, boards, enlarged from 24 to 40
Plates (21 quite new), including the latest Examples.   Together
with separate Text, also considerably enlarged, 12mo, cloth limp.
By JOHN GRANTHAM, M. Inst. C.E., &c.   2l. 2s. complete.

" Mr. Grantham's work is of great interest.   It will, we are confident, command an
extensive circulation among shipbuilders in general.   By order of the Board of Admi-
ralty, the work will form the text-book on which the examination in iron ship-building
of candidates for promotion in the dockyards will be mainly based."—*Engineering.*

Pocket-Book for Marine Engineers.

A POCKET-BOOK OF USEFUL TABLES AND FOR-
MULÆ FOR MARINE ENGINEERS.   By FRANK PROCTOR,
A.I.N.A.   Second Edition, revised and enlarged.   Royal 32mo,
leather, gilt edges, with strap, 4s.

" A most useful companion to all marine engineers."—*United Service Gazette.*
" Scarcely anything required by a naval engineer appears to have been for-
gotten."—*Iron.*

Light-Houses.

EUROPEAN LIGHT-HOUSE SYSTEMS; being a Report of a Tour of Inspection made in 1873. By Major GEORGE H. ELLIOT, Corps of Engineers, U.S.A. Illustrated by 51 Engravings and 31 Woodcuts in the Text. 8vo, 21s. cloth.

Surveying (Land and Marine).

LAND AND MARINE SURVEYING, In Reference to the Preparation of Plans for Roads and Railways, Canals, Rivers, Towns' Water Supplies, Docks and Harbours; with Description and Use of Surveying Instruments. By W. DAVIS HASKOLL, C.E. With 14 folding Plates, and numerous Woodcuts. 8vo, 12s. 6d. cloth.

"A most useful and well arranged book for the aid of a student."—*Builder*.

"Of the utmost practical utility, and may be safely recommended to all students who aspire to become clean and expert surveyors."—*Mining Journal*.

Storms.

STORMS: their Nature, Classification, and Laws, with the Means of Predicting them by their Embodiments, the Clouds. By WILLIAM BLASIUS. Crown 8vo, 10s. 6d. cloth boards.

Rudimentary Navigation.

THE SAILOR'S SEA-BOOK: a Rudimentary Treatise on Navigation. By JAMES GREENWOOD, B.A. New and enlarged edition. By W. H. ROSSER. 12mo, 3s. cloth boards.

Mathematical and Nautical Tables.

MATHEMATICAL TABLES, for Trigonometrical, Astronomical, and Nautical Calculations; to which is prefixed a Treatise on Logarithms. By HENRY LAW, C.E. Together with a Series of Tables for Navigation and Nautical Astronomy. By J. R. YOUNG, formerly Professor of Mathematics in Belfast College. New Edition. 12mo, 4s. cloth boards.

Navigation (Practical), with Tables.

PRACTICAL NAVIGATION: consisting of the Sailor's Sea-Book, by JAMES GREENWOOD and W. H. ROSSER; together with the requisite Mathematical and Nautical Tables for the Working of the Problems. By HENRY LAW, C.E., and Professor J. R. YOUNG. Illustrated with numerous Wood Engravings and Coloured Plates. 12mo, 7s. strongly half bound in leather.

WEALE'S RUDIMENTARY SERIES.

The following books in Naval Architecture, etc., are published in the above series.

MASTING, MAST-MAKING, AND RIGGING OF SHIPS. By ROBERT KIPPING, N.A. Fourteenth Edition. 12mo, 2s. 6d. cloth.

SAILS AND SAIL-MAKING. Tenth Edition, enlarged. By ROBERT KIPPING, N.A. Illustrated. 12mo, 3s. cloth boards.

NAVAL ARCHITECTURE. By JAMES PEAKE. Fourth Edition, with Plates and Diagrams. 12mo, 4s. cloth boards.

MARINE ENGINES, AND STEAM VESSELS. By ROBERT MURRAY, C.E. Seventh Edition. 12mo, 3s. 6d. cloth boards.

ARCHITECTURE, BUILDING, ETC.

Construction.

THE SCIENCE of BUILDING: An Elementary Treatise on the Principles of Construction. By E. WYNDHAM TARN, M.A., Architect. With 47 Wood Engravings. Demy 8vo. 8s. 6d. cloth.

"A very valuable book, which we strongly recommend to all students."—*Builder.*
"No architectural student should be without this hand-book."—*Architect.*

Villa Architecture.

A HANDY BOOK of VILLA ARCHITECTURE; being a Series of Designs for Villa Residences in various Styles. With Detailed Specifications and Estimates. By C. WICKES, Architect, Author of "The Spires and Towers of the Mediæval Churches of England," &c. 31 Plates, 4to, half morocco, gilt edges, 1l. 1s.

₄ Also an Enlarged edition of the above. 61 Plates, with Detailed Specifications, Estimates, &c. 2l. 2s. half morocco.

"The whole of the designs bear evidence of their being the work of an artistic architect, and they will prove very valuable and suggestive."—*Building News.*

Useful Text-Book for Architects.

THE ARCHITECT'S GUIDE: Being a Text-book of Useful Information for Architects, Engineers, Surveyors, Contractors, Clerks of Works, &c., &c. By FREDERICK ROGERS, Architect, Author of "Specifications for Practical Architecture," &c. With numerous Illustrations. Crown 8vo, 6s. cloth.

"As a text-book of useful information for architects, engineers, surveyors, &c., it would be hard to find a handier or more complete little volume."—*Standard.*

Taylor and Cresy's Rome.

THE ARCHITECTURAL ANTIQUITIES OF ROME. By the late G. L. TAYLOR, Esq., F.S.A., and EDWARD CRESY, Esq. New Edition, thoroughly revised, and supplemented under the editorial care of the Rev. ALEXANDER TAYLOR, M.A. (son of the late G. L. Taylor, Esq.), Chaplain of Gray's Inn. This is the only book which gives on a large scale, and with the precision of architectural measurement, the principal Monuments of Ancient Rome in plan, elevation, and detail. Large folio, with 130 Plates, half-bound, 3l. 3s.

₄ Originally published in two volumes, folio, at 18l. 18s.

Vitruvius' Architecture.

THE ARCHITECTURE OF MARCUS VITRUVIUS POLLIO. Translated by JOSEPH GWILT, F.S.A., F.R.A.S. Numerous Plates. 12mo, cloth limp. 5s.

The Young Architect's Book.

HINTS TO YOUNG ARCHITECTS. By GEORGE WIGHTWICK, Architect. New Edition, revised and enlarged. By G. HUSKISSON GUILLAUME, Architect. 12mo, cloth boards, 4s.

"Will be found an acquisition to pupils, and a copy ought to be considered as necessary a purchase as a box of instruments."—*Architect.*
"A large amount of information, which young architects will do well to acquire, if they wish to succeed in the everyday work of their profession."—*English Mechanic.*

Drawing for Builders and Students.

PRACTICAL RULES ON DRAWING for the OPERATIVE BUILDER and YOUNG STUDENT in ARCHITECTURE. By GEORGE PYNE. With 14 Plates, 4to, 7s. 6d. boards.

The House-Owner's Estimator.

THE HOUSE-OWNER'S ESTIMATOR; or, What will it Cost to Build, Alter, or Repair? A Price-Book adapted to the Use of Unprofessional People as well as for the Architectural Surveyor and Builder. By the late JAMES D. SIMON, A.R.I.B.A. Edited and Revised by FRANCIS T. W. MILLER, Surveyor. With numerous Illustrations. Second Edition, with the prices carefully corrected to present time. Crown 8vo, cloth, 3s. 6d.

" In two years it will repay its cost a hundred times over."—*Field.*

" A very handy book for those who want to know what a house will cost to build, alter, or repair."—*English Mechanic.*

Boiler and Factory Chimneys.

BOILER AND FACTORY CHIMNEYS ; their Draught-power and Stability, with a chapter on Lightning Conductors. By ROBERT WILSON, C.E., Author of " Treatise on Steam Boilers," &c., &c. Crown 8vo, 3s. 6d. cloth.

Civil and Ecclesiastical Building.

A BOOK ON BUILDING, CIVIL AND ECCLESIASTICAL, Including CHURCH RESTORATION. By Sir EDMUND BECKETT, Bart., LL.D., Q.C., F.R.A.S., Chancellor and Vicar-General of York. Author of " Clocks and Watches and Bells," &c. Second Edition, 12mo, 5s. cloth boards.

" A book which is always amusing and nearly always instructive. Sir E. Beckett will be read for the raciness of his style. We are able very cordially to recommend all persons to read it for themselves. The style throughout is in the highest degree condensed and epigrammatic."—*Times.*

" We commend the book to the thoughtful consideration of all who are interested in the building art."—*Builder.*

Architecture, Ancient and Modern.

RUDIMENTARY ARCHITECTURE, Ancient and Modern. Consisting of VITRUVIUS, translated by JOSEPH GWILT, F.S.A., &c., with 23 fine copper plates; GRECIAN Architecture, by the EARL of ABERDEEN; the ORDERS of Architecture, by W. H. LEEDS, Esq.; The STYLES of Architecture of Various Countries, by T. TALBOT BURY; The PRINCIPLES of DESIGN in Architecture, by E. L. GARBETT. In one volume, half-bound (pp. 1,100), copiously illustrated, 12s.

*** *Sold separately, in two vols., as follows—*

ANCIENT ARCHITECTURE. Containing Gwilt's Vitruvius and Aberdeen's Grecian Architecture. Price 6s. half-bound.

N.B.—*This is the only edition of VITRUVIUS procurable at a moderate price.*

MODERN ARCHITECTURE. Containing the Orders, by Leeds ; The Styles, by Bury; and Design, by Garbett. 6s. half-bound.

House Painting.

HOUSE PAINTING, GRAINING, MARBLING, AND SIGN WRITING : a Practical Manual of. With 9 Coloured Plates of Woods and Marbles, and nearly 150 Wood Engravings. By ELLIS A. DAVIDSON, Author of " Building Construction," &c. Third Edition, carefully revised. 12mo, 6s. cloth boards.

" Contains a mass of information of use to the amateur and of value to the practical man."—*English Mechanic.*

Plumbing.

PLUMBING ; a Text-book to the Practice of the Art or Craft of the Plumber. With chapters upon House-drainage, embodying the latest Improvements. By W. P. BUCHAN, Sanitary Engineer. Second Edition, enlarged, with 300 illustrations, 12mo.   4s. cloth.
"The chapters on house-drainage may be usefully consulted, not only by plumbers, but also by engineers and all engaged or interested in house-building."—*Iron.*

Handbook of Specifications.

THE HANDBOOK OF SPECIFICATIONS ; or, Practical Guide to the Architect, Engineer, Surveyor, and Builder, in drawing up Specifications and Contracts for Works and Constructions. Illustrated by Precedents of Buildings actually executed by eminent Architects and Engineers. By Professor THOMAS L. DONALD-SON, M.I.B.A. New Edition, in One large volume, 8vo, with upwards of 1000 pages of text, and 33 Plates, cloth, 1l. 11s. 6d.
"In this work forty-four specifications of executed works are given. . . . Donald-son's Handbook of Specifications must be bought by all architects."—*Builder.*

Specifications for Practical Architecture.

SPECIFICATIONS FOR PRACTICAL ARCHITECTURE: A Guide to the Architect, Engineer, Surveyor, and Builder ; with an Essay on the Structure and Science of Modern Buildings. By FREDERICK ROGERS, Architect. 8vo, 15s. cloth.
*** A volume of specifications of a practical character being greatly required, and the old standard work of Alfred Bartholomew being out of print, the author, on the basis of that work, has produced the above.—*Extract from Preface.*

Designing, Measuring, and Valuing.

THE STUDENT'S GUIDE to the PRACTICE of MEA-SURING and VALUING ARTIFICERS' WORKS ; containing Directions for taking Dimensions, Abstracting the same, and bringing the Quantities into Bill, with Tables of Constants, and copious Memoranda for the Valuation of Labour and Materials in the respective Trades of Bricklayer and Slater, Carpenter and Joiner, Painter and Glazier, Paperhanger, &c. With 43 Plates and Wood-cuts. Originally edited by EDWARD DOBSON, Architect. New Edition, re-written, with Additions on Mensuration and Construction, and useful Tables for facilitating Calculations and Measurements. By E. WYNDHAM TARN, M.A., 8vo, 10s. 6d. cloth.
"Well fulfils the promise of its title-page. Mr. Tarn's additions and revisions have much increased the usefulness of the work."—*Engineering.*

Beaton's Pocket Estimator.

THE POCKET ESTIMATOR FOR THE BUILDING TRADES, being an easy method of estimating the various parts of a Building collectively, more especially applied to Carpenters' and Joiners' work, priced according to the present value of material and labour. By A. C. BEATON, Author of "Quantities and Measurements." Second Edition. Waistcoat-pocket size. 1s. 6d.

Beaton's Builders' and Surveyors' Technical Guide.

THE POCKET TECHNICAL GUIDE AND MEASURER FOR BUILDERS AND SURVEYORS: containing a Complete Explanation of the Terms used in Building Construction, Memoranda for Reference, Technical Directions for Measuring Work in all the Building Trades, &c. By A. C. BEATON. Second Edit. Waistcoat-pocket size. 1s. 6d.

Builder's and Contractor's Price Book.

LOCKWOOD & CO.'S BUILDER'S AND CONTRACTOR'S PRICE BOOK, containing the latest prices of all kinds of Builders' Materials and Labour, and of all Trades connected with Building, &c., &c. The whole revised and edited by F. T. W. MILLER, Architect and Surveyor. Fcap. half-bound, 4s.

CARPENTRY, TIMBER, ETC.

Tredgold's Carpentry, new and cheaper Edition.

THE ELEMENTARY PRINCIPLES OF CARPENTRY : a Treatise on the Pressure and Equilibrium of Timber Framing, the Resistance of Timber, and the Construction of Floors, Arches, Bridges, Roofs, Uniting Iron and Stone with Timber, &c. To which is added an Essay on the Nature and Properties of Timber, &c., with Descriptions of the Kinds of Wood used in Building ; also numerous Tables of the Scantlings of Timber for different purposes, the Specific Gravities of Materials, &c. By THOMAS TREDGOLD, C.E. Edited by PETER BARLOW, F.R.S. Fifth Edition, corrected and enlarged. With 64 Plates (11 of which now first appear in this edition), Portrait of the Author, and several Woodcuts. In 1 vol., 4to, published at 2l. 2s., reduced to 1l. 5s. cloth.

"Ought to be in every architect's and every builder's library, and those who do not already possess it ought to avail themselves of the new issue."—*Builder.*

"A work whose monumental excellence must commend it wherever skilful carpentry is concerned. The Author's principles are rather confirmed than impaired by time. The additional plates are of great intrinsic value."—*Building News.*

Grandy's Timber Tables.

THE TIMBER IMPORTER'S, TIMBER MERCHANT'S, and BUILDER'S STANDARD GUIDE. By RICHARD E. GRANDY. Comprising :—An Analysis of Deal Standards, Home and Foreign, with comparative Values and Tabular Arrangements for Fixing Nett Landed Cost on Baltic and North American Deals, including all intermediate Expenses, Freight, Insurance, &c., &c. ; together with Copious Information for the Retailer and Builder. 2nd Edition. Carefully revised and corrected. 12mo, 3s. 6d. cloth.

"Everything it pretends to be : built up gradually, it leads one from a forest to a treenail, and throws in, as a makeweight, a host of material concerning bricks, columns, cisterns, &c.—all that the class to whom it appeals requires."—*English Mechanic.*

Timber Freight Book.

THE TIMBER IMPORTERS' AND SHIPOWNERS' FREIGHT BOOK : Being a Comprehensive Series of Tables for the Use of Timber Importers, Captains of Ships, Shipbrokers, Builders, and all Dealers in Wood whatsoever. By WILLIAM RICHARDSON, Timber Broker. Crown 8vo, 6s. cloth.

Tables for Packing-Case Makers.

PACKING-CASE TABLES ; showing the number of Superficial Feet in Boxes or Packing-Cases, from six inches square and upwards. Compiled by WILLIAM RICHARDSON, Accountant. Second Edition. Oblong 4to, 3s. 6d. cloth.

"Will save much labour and calculation to packing-case makers and those who use packing-cases."—*Grocer.*   "Invaluable labour-saving tables."—*Ironmonger.*

Horton's Measurer.

THE COMPLETE MEASURER; setting forth the Measurement of Boards, Glass, &c.; Unequal-sided, Square-sided, Octagonal-sided, Round Timber and Stone, and Standing Timber. With just allowances for the bark in the respective species of trees, and proper deductions for the waste in hewing the trees, &c.; also a Table showing the solidity of hewn or eight-sided timber, or of any octagonal-sided column. By RICHARD HORTON. Fourth edition, with considerable and valuable additions, 12mo, strongly bound in leather, 5s.

Horton's Underwood and Woodland Tables.

TABLES FOR PLANTING AND VALUING UNDERWOOD AND WOODLAND; also Lineal, Superficial, Cubical, and Decimal Tables, &c. By R. HORTON. 12mo, 2s. leather.

Nicholson's Carpenter's Guide.

THE CARPENTER'S NEW GUIDE; or, BOOK of LINES for CARPENTERS: comprising all the Elementary Principles essential for acquiring a knowledge of Carpentry. Founded on the late PETER NICHOLSON'S standard work. A new Edition, revised by ARTHUR ASHPITEL, F.S.A., together with Practical Rules on Drawing, by GEORGE PYNE. With 74 Plates, 4to, 1l. 1s. cloth.

Dowsing's Timber Merchant's Companion.

THE TIMBER MERCHANT'S AND BUILDER'S COMPANION; containing New and Copious Tables of the Reduced Weight and Measurement of Deals and Battens, of all sizes, from One to a Thousand Pieces, also the relative Price that each size bears per Lineal Foot to any given Price per Petersburgh Standard Hundred, &c., &c. Also a variety of other valuable information. By WILLIAM DOWSING, Timber Merchant. Third Edition, Revised. Crown 8vo, 3s. cloth.
"Everything is as concise and clear as it can possibly be made. There can be no doubt that every timber merchant and builder ought to possess it."—*Hull Advertiser.*

Practical Timber Merchant.

THE PRACTICAL TIMBER MERCHANT, being a Guide for the use of Building Contractors, Surveyors, Builders, &c., comprising useful Tables for all purposes connected with the Timber Trade, Essay on the Strength of Timber, Remarks on the Growth of Timber, &c. By W. RICHARDSON. Fcap. 8vo, 3s. 6d. cl.

Woodworking Machinery.

WOODWORKING MACHINERY; its Rise, Progress, and Construction. With Hints on the Management of Saw Mills and the Economical Conversion of Timber. Illustrated with Examples of Recent Designs by leading English, French, and American Engineers. By M. POWIS BALE, M.I.M.E. Large crown 8vo, 12s. 6d. cloth.   [*Just published.*
"Mr. Bale is evidently an expert on the subject, and he has collected so much information that his book is all-sufficient for builders and others engaged in the conversion of timber."—*Architect.*
"The most comprehensive compendium of wood-working machinery we have seen. The author is a thorough master of his subject."—*Building News.*
"It should be in the office of every wood-working factory."—*English Mechanic.*

MECHANICS, ETC.

Mechanic's Workshop Companion.

THE OPERATIVE MECHANIC'S WORKSHOP COM-PANION, and THE SCIENTIFIC GENTLEMAN'S PRAC-TICAL ASSISTANT. By W. TEMPLETON. 12th Edit., with Mechanical Tables for Operative Smiths, Millwrights, Engineers, &c. ; and an Extensive Table of Powers and Roots, 12mo, 5s. bound.

" As a text-book in which mechanical and commercial demands are judiciously met, TEMPLETON'S COMPANION stands unrivalled."—*Mechanics' Magazine.*

" Admirably adapted to the wants of a very large class. It has met with great success in the engineering workshop, as we can testify ; and there are a great many men who, in a great measure, owe their rise in life to this little work."—*Building News.*

Engineer's and Machinist's Assistant.

THE ENGINEER'S, MILLWRIGHT'S, and MACHINIST'S PRACTICAL ASSISTANT ; comprising a Collection of Useful Tables, Rules, and Data. Compiled and Arranged, with Original Matter, by WM. TEMPLETON. 6th Edition. 18mo, 2s. 6d. cloth.

" A more suitable present to an apprentice to any of the mechanical trades could not possibly be made."—*Building News.*

Superficial Measurement.

THE TRADESMAN'S GUIDE TO SUPERFICIAL MEA-SUREMENT. Tables calculated from 1 to 200 inches in length, by 1 to 108 inches in breadth. For the use of Architects, Engineers, Timber Merchants, Builders, &c. By J. HAWKINGS. Fcp. 3s. 6d. cl.

The High-Pressure Steam Engine.

THE HIGH-PRESSURE STEAM ENGINE ; an Exposition of its Comparative Merits, and an Essay towards an Improved System of Construction, adapted especially to secure Safety and Economy. By Dr. ERNST ALBAN, Practical Machine Maker, Plau, Mecklenberg. Translated from the German, with Notes, by Dr. POLE, F.R.S., M.I.C.E., &c. With 28 Plates, 8vo, 16s. 6d. cl.

Steam Boilers.

A TREATISE ON STEAM BOILERS : their Strength, Con-struction, and Economical Working. By R. WILSON, C.E. Fifth Edition. 12mo, 6s. cloth.

" The best work on boilers which has come under our notice."—*Engineering.*

" The best treatise that has ever been published on steam boilers."—*Engineer.*

Power in Motion.

POWER IN MOTION : Horse Power, Toothed Wheel Gearing, Long and Short Driving Bands, Angular Forces, &c. By JAMES ARMOUR, C.E. With 73 Diagrams. 12mo, 3s., cloth.

Mechanics.

THE HANDBOOK OF MECHANICS. By DIONYSIUS LARDNER, D.C.L., formerly Professor of Natural Philosophy and Astronomy in University College, London. New Edition, Edited and considerably Enlarged, by BENJAMIN LOEWY, F.R.A.S., &c., &c. With 378 Illustrations, post 8vo, 6s. cloth.

" The explanations throughout are studiously popular, and care has been taken to show the application of the various branches of physics to the industrial arts, and to the practical business of life."—*Mining Journal.*

MATHEMATICS, TABLES, ETC.

Gregory's Practical Mathematics.

MATHEMATICS for PRACTICAL MEN ; being a Common-place Book of Pure and Mixed Mathematics. Designed chiefly for the Use of Civil Engineers, Architects, and Surveyors. Part I. PURE MATHEMATICS—comprising Arithmetic, Algebra, Geometry, Mensuration, Trigonometry, Conic Sections, Properties of Curves. Part II. MIXED MATHEMATICS—comprising Mechanics in general, Statics, Dynamics, Hydrostatics, Hydrodynamics, Pneumatics, Mechanical Agents, Strength of Materials. With an Appendix of copious Logarithmic and other Tables. By OLINTHUS GREGORY, LL.D., F.R.A.S. Enlarged by HENRY LAW, C.E. 4th Edition, carefully revised and corrected by J. R. YOUNG, formerly Profes-sor of Mathematics, Belfast Coll. With 13 Plates. 8vo, 1l. 1s. cloth.

"The engineer or architect will here find ready to his hand, rules for solving nearly every mathematical difficulty that may arise in his practice. The rules are in all cases explained by means of examples clearly worked out."—*Builder.*

"One of the most serviceable books for practical mechanics. . . . "—*Building News.*

The Metric System.

A SERIES OF METRIC TABLES, in which the British Standard Measures and Weights are compared with those of the Metric System at present in use on the Continent. By C. H. DOWLING, C.E. 2nd Edit., revised and enlarged. 8vo, 10s. 6d. cl.

"Their accuracy has been certified by Prof. Airy, Astronomer-Royal."—*Builder.*

Inwood's Tables, greatly enlarged and improved.

TABLES FOR THE PURCHASING of ESTATES, Freehold, Copyhold, or Leasehold; Annuities, Advowsons, &c., and for the Renewing of Leases held under Cathedral Churches, Colleges, or other corporate bodies; for Terms of Years certain, and for Lives ; also for Valuing Reversionary Estates, Deferred Annuities, Next Presentations, &c., together with Smart's Five Tables of Compound Interest, and an Extension of the same to Lower and Intermediate Rates. By WILLIAM INWOOD, Architect. The 21st edition, with considerable additions, and new and valuable Tables of Logarithms for the more Difficult Computations of the Interest of Money, Dis-count, Annuities, &c., by M. FÉDOR THOMAN, of the Société Crédit Mobilier of Paris. 12mo, 8s. cloth.

"Those interested in the purchase and sale of estates, and in the adjustment of compensation cases. as well as in transactions in annuities, life insurances, &c., will find the present editon of eminent service."—*Engineering.*

Geometry for the Architect, Engineer, &c.

PRACTICAL GEOMETRY, for the Architect, Engineer, and Mechanic ; giving Rules for the Delineation and Application of various Geometrical Lines, Figures and Curves. By E. W. TARN, M.A., Architect, Author of "The Science of Building," &c. With 164 Illustrations. Demy 8vo. 12s. 6d. cloth.

Mathematical Instruments.

MATHEMATICAL INSTRUMENTS : Their Construction, Adjustment, Testing, and Use ; comprising Drawing, Measuring, Optical, Surveying, and Astronomical Instruments. By J. F. HEATHER, M.A. Enlarged Edition, for the most part entirely re-written. Numerous Woodcuts. 12mo, 5s. cloth.

Compound Interest and Annuities.

THEORY of COMPOUND INTEREST and ANNUITIES; with Tables of Logarithms for the more Difficult Computations of Interest, Discount, Annuities, &c., in all their Applications and Uses for Mercantile and State Purposes. By Fédor Thoman, of the Société Crédit Mobilier, Paris. 3rd Edit., 12mo, 4s. 6d. cl.

"A very powerful work, and the Author has a very remarkable command of his subject."—*Professor A. de Morgan.*

Iron and Metal Trades' Calculator.

THE IRON AND METAL TRADES' COMPANION: Being a Calculator containing a Series of Tables upon a new and comprehensive plan for expeditiously ascertaining the value of any goods bought or sold by weight, from 1s. per cwt. to 112s. per cwt., and from one farthing per lb. to 1s. per lb. Each Table extends from one lb. to 100 tons. By T. Downie. 396 pp., 9s., leather.

" A most useful set of tables, and will supply a want, for nothing like them before existed."—*Building News.*

Iron and Steel.

'IRON AND STEEL': a Work for the Forge, Foundry, Factory, and Office. Containing Information for Ironmasters and their Stocktakers; Managers of Bar, Rail, Plate, and Sheet Rolling Mills; Iron and Metal Founders; Iron Ship and Bridge Builders; Mechanical, Mining, and Consulting Engineers; Architects, Builders, &c. By Charles Hoare, Author of 'The Slide Rule,' &c. Eighth Edition. With folding Scales of "Foreign Measures compared with the English Foot," and "fixed Scales of Squares, Cubes, and Roots, Areas, Decimal Equivalents, &c." Oblong, 32mo, 6s., leather, elastic-band.

" For comprehensiveness the book has not its equal."—*Iron.*

Comprehensive Weight Calculator.

THE WEIGHT CALCULATOR; being a Series of Tables upon a New and Comprehensive Plan, exhibiting at one Reference the exact Value of any Weight from 1 lb. to 15 tons, at 300 Progressive Rates, from 1 Penny to 168 Shillings per cwt., and containing 186,000 Direct Answers, which, with their Combinations, consisting of a single addition (mostly to be performed at sight), will afford an aggregate of 10,266,000 Answers; the whole being calculated and designed to ensure Correctness and promote Despatch. By Henry Harben, Accountant, Sheffield. New Edition. Royal 8vo, 1l. 5s., strongly half-bound.

Comprehensive Discount Guide.

THE DISCOUNT GUIDE: comprising several Series of Tables for the use of Merchants, Manufacturers, Ironmongers, and others, by which may be ascertained the exact profit arising from any mode of using Discounts, either in the Purchase or Sale of Goods, and the method of either Altering a Rate of Discount, or Advancing a Price, so as to produce, by one operation, a sum that will realise any required profit after allowing one or more Discounts: to which are added Tables of Profit or Advance from $1\frac{1}{4}$ to 90 per cent., Tables of Discount from $1\frac{1}{4}$ to $98\frac{3}{4}$ per cent., and Tables of Commission, &c., from $\frac{1}{8}$ to 10 per cent. By Henry Harben, Accountant. New Edition, Demy 8vo. £1 5s., half-bound.

SCIENCE AND ART.

Dentistry.

MECHANICAL DENTISTRY. A Practical Treatise on the Construction of the various kinds of Artificial Dentures. Comprising also Useful Formulæ, Tables, and Receipts for Gold Plate, Clasps, Solders, etc., etc. By CHARLES HUNTER. With numerous Wood Engravings. Crown 8vo, 7s. 6d. cloth.

"The work is very practical."—*Monthly Review of Dental Surgery.*
"An authoritative treatise . . . . . We can strongly recommend Mr. Hunter's treatise to all students preparing for the profession of dentistry, as well as to every mechanical dentist."—*Dublin Journal of Medical Science.*
"The best book on the subject with which we are acquainted."—*Medical Press and Circular.*

Brewing.

A HANDBOOK FOR YOUNG BREWERS. By HERBERT EDWARDS WRIGHT, B.A. Crown 8vo, 3s. 6d. cloth.

"A thoroughly scientific treatise in popular language. It is evident that the author has mastered his subject in its scientific aspects."—*Morning Advertiser.*
"We would particularly recommend teachers of the art to place it in every pupil's hands, and we feel sure its perusal will be attended with advantage."—*Brewer.*

Gold and Gold-Working.

THE GOLDSMITH'S HANDBOOK : containing full instructions for the Alloying and Working of Gold. Including the Art of Alloying, Melting, Reducing, Colouring, Collecting and Refining. The processes of Manipulation, Recovery of Waste, Chemical and Physical Properties of Gold, with a new System of Mixing its Alloys ; Solders, Enamels, and other useful Rules and Recipes, &c. By GEORGE E. GEE, Goldsmith and Silversmith. Second Edition, considerably enlarged. 12mo, 3s. 6d. cloth boards.

"A good, sound, technical educator, and will be generally accepted as an authority. It gives full particulars for mixing alloys and enamels, is essentially a book for the workshop, and exactly fulfils the purpose intended."—*Horological Journal.*
"The best work yet printed on its subject for a reasonable price. We have no doubt that it will speedily become a standard book which few will care to be without."—*Jeweller and Metalworker.*
"We consider that the trade owes not a little to Mr. Gee, who has in two volumes compressed almost the whole of its literature, and we doubt not that many a young beginner will owe a part of his future success to a diligent study of the pages which are peculiarly well adapted to his use."—*Clerkenwell Press.*
"It is essentially a practical manual, intended primarily for the use of working jewellers, but is well adapted to the wants of amateurs and apprentices, containing, as it does, trustworthy information that only a practical man can supply."—*English Mechanic.*

Silver and Silver Working.

THE SILVERSMITH'S HANDBOOK, containing full Instructions for the Alloying and Working of Silver, including the different modes of refining and melting the metal, its solders, the preparation of imitation alloys, methods of manipulation, prevention of waste, instructions for improving and finishing the surface of the work, together with other useful information and memoranda. By GEORGE E. GEE, Jeweller, &c. 12mo, 3s. 6d. cloth boards.

"This work is destined to take up as good a position in technical literature as the *Practical Goldworker*, a book which has passed through the ordeal of critical examination and business tests with great success."—*Jeweller and Metalworker.*
"The chief merit of the work is its practical character. The workers in the trade will speedily discover its merits when they sit down to study it."—*English Mechanic.*
"This work forms a valuable sequel to the author's *Practical Goldworker*, and supplies a want long felt in the silver trade."—*Silversmith's Trade Journal.*

Electric Lighting.

ELECTRIC LIGHT : Its Production and Use, embodying plain Directions for the Working of Galvanic Batteries, Electric Lamps, and Dynamo-Electric Machines. By J. W. URQUHART, C. E., Author of "Electroplating : a Practical Handbook." Edited by F. C. WEBB, M.I.C.E., M.S.T.E., With 94 Illustrations. Crown 8vo, 7s. 6d. cloth.   [*Just published.*

"It is the only work at present available, which gives in language intelligible for the most part to the ordinary reader, a general but concise history of the means which have been adopted up to the present time in producing the electric light."—*Metropolitan.*

"An important addition to the literature of the electric light. Students of the subject should not fail to read it."—*Colliery Guardian.*

"As a popular and practical treatise on the subject, the volume may be thoroughly recommended."—*Bristol Mercury.*

Electroplating, etc.

ELECTROPLATING : A Practical Handbook, including the Practice of Electrotyping. By J. W. URQUHART, C.E. With numerous Illustrations. Crown 8vo, 5s. cloth.

"The volume is without a rival in its particular sphere, and the lucid style in which it is written commends it to those amateurs and experimental electrotypers who have but slight, if any, knowledge of the processes of the art to which they turn their attention."—*Design and Work.*

"A large amount of thoroughly practical information."—*Telegraphic Journal.*

"An excellent practical manual."—*Engineering.*

"The information given appears to be based on direct personal knowledge. . . . Its science is sound, and the style is always clear." —*Athenæum.*

"Any ordinarily intelligent person may become an adept in electro-deposition with a very little science indeed, and this is the book to show him or her the way." —*Builder.*

The Military Sciences.

AIDE-MÉMOIRE to the MILITARY SCIENCES. Framed from Contributions of Officers and others connected with the different Services. Originally edited by a Committee of the Corps of Royal Engineers. Second Edition, most carefully revised by an Officer of the Corps, with many additions ; containing nearly 350 Engravings and many hundred Woodcuts. 3 vols. royal 8vo, extra cloth boards, and lettered, 4l. 10s.

"A compendious encyclopædia of military knowledge."—*Edinburgh Review.*

"The most comprehensive work of reference to the military and collateral sciences." —*Volunteer Service Gazette.*

Field Fortification.

A TREATISE on FIELD FORTIFICATION, the ATTACK of FORTRESSES, MILITARY MINING, and RECON-NOITRING. By Colonel I. S. MACAULAY, late Professor of Fortification in the R. M. A., Woolwich. Sixth Edition, crown 8vo, cloth, with separate Atlas of 12 Plates, 12s. complete.

Dye-Wares and Colours.

THE MANUAL of COLOURS and DYE-WARES : their Properties, Applications, Valuation, Impurities, and Sophistications. For the Use of Dyers, Printers, Drysalters, Brokers, &c. By J. W. SLATER. Post 8vo, 7s. 6d. cloth.

"A complete encyclopædia of the *materia tinctoria*. The information is full nd precise, and the methods of determining the value of articles liable to sophistica-on, are practical as well as valuable."—*Chemist and Druggist.*

The Alkali Trade—Sulphuric Acid, etc.

A MANUAL OF THE ALKALI TRADE, including the Manufacture of Sulphuric Acid, Sulphate of Soda, and Bleaching Powder. By JOHN LOMAS, Alkali Manufacturer, Newcastle-upon-Tyne and London. With 232 Illustrations and Working Drawings, and containing 386 pages of text. Super-royal 8vo, 2l 12s. 6d. cloth.   [Just published.

This work provides (1) a Complete Handbook for intending Alkali and Sulphuric Acid Manufacturers, and for those already in the field who desire to improve their plant, or to become practically acquainted with the latest processes and developments of the trade ; (2) a Handy Volume which Manufacturers can put into the hands of their Managers and Foremen as a useful guide in their daily rounds of duty.

SYNOPSIS OF CONTENTS.

Chap. I. Choice of Site and General Plan of Works—II. Sulphuric Acid—III. Recovery of the Nitrogen Compounds, and Treatment of Small Pyrites—IV. The Salt Cake Process—V. Legislation upon the Noxious Vapours Question—VI. The Hargreaves' and Jones' Processes—VII. The Balling Process—VIII. Lixiviation and Salting Down—IX. Carbonating or Finishing—X. Soda Crystals — XI. Refined Alkali — XII. Caustic Soda — XIII. Bi-carbonate of Soda — XIV. Bleaching Powder — XV. Utilisation of Tank Waste—XVI. General Remarks—Four Appendices, treating of Yields, Sulphuric Acid Calculations, Anemometers, and Foreign Legislation upon the Noxious Vapours Question.

"The author has given the fullest, most practical, and, to all concerned in the alkali trade, most valuable mass of information that, to our knowledge, has been published in any language."—*Engineer.*

"This book is written by a manufacturer for manufacturers. The working details of the most approved forms of apparatus are given, and these are accompanied by no less than 232 wood engravings, all of which may be used for the purposes of construction. Every step in the manufacture is very fully described in this manual, and each improvement explained. Everything which tends to introduce economy into the technical details of this trade receives the fullest attention. The book has been produced with great completeness."—*Athenæum.*

"The author is not one of those clever compilers who, on short notice, will 'read up' any conceivable subject, but a practical man in the best sense of the word. We find here not merely a sound and luminous explanation of the chemical principles of the trade, but a notice of numerous matters which have a most important bearing on the successful conduct of alkali works, but which are generally overlooked by even the most experienced technological authors. This most valuable book, which we trust will be generally appreciated, we must pronounce a credit alike to its author and to the enterprising firm who have undertaken its publication."—*Chemical Review.*

Chemical Analysis.

THE COMMERCIAL HANDBOOK of CHEMICAL ANALYSIS ; or Practical Instructions for the determination of the Intrinsic or Commercial Value of Substances used in Manufactures, in Trades, and in the Arts. By A. NORMANDY, Author of "Practical Introduction to Rose's Chemistry," and Editor of Rose's "Treatise on Chemical Analysis." *New Edition.* Enlarged, and to a great extent re-written, by HENRY M. NOAD, Ph. D., F.R.S. With numerous Illustrations. Cr. 8vo, 12s. 6d. cloth.

"We recommend this book to the careful perusal of every one ; it may be truly affirmed to be of universal interest, and we strongly recommend it to our readers as a guide, alike indispensable to the housewife as to the pharmaceutical practitioner."—*Medical Times.*

"Essential to the analysts appointed under the new Act. The most recent result are given, and the work is well edited and carefully written."—*Nature.*

Dr. Lardner's Museum of Science and Art.

THE MUSEUM OF SCIENCE AND ART. Edited by
DIONYSIUS LARDNER, D.C.L., formerly Professor of Natural Philosophy and Astronomy in University College, London. With upwards of 1200 Engravings on Wood. In 6 Double Volumes.
Price £1 1s., in a new and-elegant cloth binding, or handsomely
bound in half morocco, 31s. 6d.

OPINIONS OF THE PRESS.

"This series, besides affording popular but sound instruction on scientific subjects,
with which the humblest man in the country ought to be acquainted, also undertakes
that teaching of 'common things' which every well-wisher of his kind is anxious to
promote. Many thousand copies of this serviceable publication have been printed,
in the belief and hope that the desire for instruction and improvement widely prevails; and we have no fear that such enlightened faith will meet with disappoint-
ment."—*Times.*

"A cheap and interesting publication, alike informing and attractive. The papers
combine subjects of importance and great scientific knowledge, considerable induc-
tive powers, and a popular style of treatment."—*Spectator.*

"The 'Museum of Science and Art' is the most valuable contribution that has
ever been made to the Scientific Instruction of every class of society."—*Sir David
Brewster in the North British Review.*

"Whether we consider the liberality and beauty of the illustrations, the charm of
the writing, or the durable interest of the matter, we must express our belief that
there is hardly to be found among the new books, one that would be welcomed by
people of so many ages and classes as a valuable present."—*Examiner.*

*** *Separate books formed from the above, suitable for Workmen's
Libraries, Science Classes, &c.*

COMMON THINGS EXPLAINED. Containing Air, Earth, Fire,
Water, Time, Man, the Eye, Locomotion, Colour, Clocks and
Watches, &c. 233 Illustrations, cloth gilt, 5s.

THE MICROSCOPE. Containing Optical Images, Magnifying
Glasses, Origin and Description of the Microscope, Microscopic
Objects, the Solar Microscope, Microscopic Drawing and Engrav-
ing, &c. 147 Illustrations, cloth gilt, 2s.

POPULAR GEOLOGY. Containing Earthquakes and Volcanoes,
the Crust of the Earth, etc. 201 Illustrations, cloth gilt, 2s. 6d.

POPULAR PHYSICS. Containing Magnitude and Minuteness, the
Atmosphere, Meteoric Stones, Popular Fallacies, Weather Prog-
nostics, the Thermometer, the Barometer, Sound, &c. 85 Illustrations, cloth gilt, 2s. 6d.

STEAM AND ITS USES. Including the Steam Engine, the Locomotive, and Steam Navigation. 89 Illustrations, cloth gilt, 2s.

POPULAR ASTRONOMY. Containing How to Observe the
Heavens. The Earth, Sun, Moon, Planets. Light, Comets,
Eclipses, Astronomical Influences, &c. 182 Illustrations, 4s. 6d.

THE BEE AND WHITE ANTS: Their Manners and Habits.
With Illustrations of Animal Instinct and Intelligence. 135 Illustrations, cloth gilt, 2s.

THE ELECTRIC TELEGRAPH POPULARISED. To render
intelligible to all who can Read, irrespective of any previous Scientific Acquirements, the various forms of Telegraphy in Actual
Operation. 100 Illustrations, cloth gilt, 1s. 6d.

Dr. Lardner's Handbooks of Natural Philosophy.

** The following five volumes, though each is Complete in itself, and to be purchased separately, form A COMPLETE COURSE OF NATURAL PHILOSOPHY, and are intended for the general reader who desires to attain accurate knowledge of the various departments of Physical Science, without pursuing them according to the more profound methods of mathematical investigation. The style is studiously popular. It has been the author's aim to supply Manuals such as are required by the Student, the Engineer, the Artisan, and the superior classes in Schools.

THE HANDBOOK OF MECHANICS. Enlarged and almost rewritten by BENJAMIN LOEWY, F.R.A.S. With 378 Illustrations. Post 8vo, 6s. cloth.

"The perspicuity of the original has been retained, and chapters which had become obsolete, have been replaced by others of more modern character. The explanations throughout are studiously popular, and care has been taken to show the application of the various branches of physics to the industrial arts, and to the practical business of life."—*Mining Journal.*

THE HANDBOOK of HYDROSTATICS and PNEUMATICS. New Edition, Revised and Enlarged by BENJAMIN LOEWY, F.R.A.S. With 236 Illustrations. Post 8vo, 5s. cloth.

" For those 'who desire to attain an accurate knowledge of physical science without the profound methods of mathematical investigation,' this work is not merely intended, but well adapted."—*Chemical News.*

THE HANDBOOK OF HEAT. Edited and almost entirely Rewritten by BENJAMIN LOEWY, F.R.A.S., etc. 117 Illustrations. Post 8vo, 6s. cloth.

"The style is always clear and precise, and conveys instruction without leaving any cloudiness or lurking doubts behind."—*Engineering.*

THE HANDBOOK OF OPTICS. New Edition. Edited by T. OLVER HARDING, B.A. 298 Illustrations. Post 8vo, 5s. cloth.

" Written by one of the ablest English scientific writers, beautifully and elaborately illustrated."—*Mechanics' Magazine.*

THE HANDBOOK OF ELECTRICITY, MAGNETISM, and ACOUSTICS. New Edition. Edited by GEO. CAREY FOSTER, B.A., F.C.S. With 400 Illustrations. Post 8vo, 5s. cloth.

" The book could not have been entrusted to any one better calculated to preserve the terse and lucid style of Lardner, while correcting his errors and bringing up his work to the present state of scientific knowledge."—*Popular Science Review.*

Dr. Lardner's Handbook of Astronomy.

THE HANDBOOK OF ASTRONOMY. Forming a Companion to the "Handbooks of Natural Philosophy." By DIONYSIUS LARDNER, D.C.L., formerly Professor of Natural Philosophy and Astronomy in University College, London. Fourth Edition. Revised and Edited by EDWIN DUNKIN, F.R.S., Royal Observatory, Greenwich. With 38 Plates and upwards of 100 Woodcuts. In 1 vol., small 8vo, 550 pages, 9s. 6d., cloth.

" Probably no other book contains the same amount of information in so compendious and well-arranged a form—certainly none at the price at which this is offered to the public."—*Athenæum.*

" We can do no other than pronounce this work a most valuable manual of astronomy, and we strongly recommend it to all who wish to acquire a general—but at the same time correct—acquaintance with this sublime science."— *Quarterly Journal of Science.*

Dr. Lardner's Handbook of Animal Physics.

THE HANDBOOK OF ANIMAL PHYSICS. By DR. LARDNER. With 520 Illustrations. New edition, small 8vo, cloth, 732 pages, 7s. 6d.

" We have no hesitation in cordially recommending it."—*Educational Times.*

Dr. Lardner's School Handbooks.

NATURAL PHILOSOPHY FOR SCHOOLS. By Dr. Lardner. 328 Illustrations. Sixth Edition. 1 vol. 3s. 6d. cloth.

"Conveys, in clear and precise terms, general notions of all the principal divisions of Physical Science."—*British Quarterly Review.*

ANIMAL PHYSIOLOGY FOR SCHOOLS. By Dr. Lardner. With 190 Illustrations. Second Edition. 1 vol. 3s. 6d. cloth.

"Clearly written, well arranged, and excellently illustrated."—*Gardeners' Chronicle.*

Dr. Lardner's Electric Telegraph.

THE ELECTRIC TELEGRAPH. By Dr. Lardner. New Edition. Revised and Re-written, by E. B. Bright, F.R.A.S. 140 Illustrations. Small 8vo, 2s. 6d. cloth.

"One of the most readable books extant on the Electric Telegraph."—*Eng. Mechanic.*

Electricity.

A MANUAL of ELECTRICITY; including Galvanism, Magnetism, Diamagnetism, Electro-Dynamics, Magneto-Electricity, and the Electric Telegraph. By Henry M. Noad, Ph.D., F.C.S. Fourth Edition, with 500 Woodcuts. 8vo, 1l. 4s. cloth.

"The accounts given of electricity and galvanism are not only complete in a scientific sense, but, which is a rarer thing, are popular and interesting."—*Lancet.*

Text-Book of Electricity.

THE STUDENT'S TEXT-BOOK OF ELECTRICITY. By Henry M. Noad, Ph.D., F.R.S., F.C.S. New Edition, carefully Revised. With an Introduction and Additional Chapters by W. H. Preece, M.I.C.E., Vice-President of the Society of Telegraph Engineers, &c. With 470 Illustrations. Crown 8vo, 12s. 6d. cloth. [*Just published.*

"A reflex of the existing state of Electrical Science adapted for students."—W. H. Preece, Esq., vide "Introduction."

"We can recommend Dr. Noad's book for clear style, great range of subject, a good index, and a plethora of woodcuts. Such collections as the present are indispensable."—*Athenæum.*

"An admirable text-book for every student—beginner or advanced—of electricity."—*Engineering.*

"A most elaborate compilation of the facts of electricity and magnetism."—*Popular Science Review.*

"May be recommended to students as one of the best text-books on the subject that they can have. . . . Mr. Preece appears to have introduced all the newest inventions in the shape of telegraphic, telephonic, and electric-lighting apparatus."—*English Mechanic.*

"The work contains everything that the student can require, it is well illustrated, clearly written, and possesses a good index."—*Academy.*

"One of the best and most useful compendiums of any branch of science in our literature."—*Iron.*

"Under the editorial hand of Mr. Preece the late Dr. Noad's text-book of electricity has grown into an admirable handbook."—*Westminster Review.*

Geology and Genesis.

THE TWIN RECORDS OF CREATION; or, Geology and Genesis, their Perfect Harmony and Wonderful Concord. By George W. Victor le Vaux. Numerous Illustrations. Fcap. 8vo, 5s. cloth.

"A valuable contribution to the evidences of revelation. and disposes very conclusively of the arguments of those who would set God's Works against God's Word. No real difficulty is shirked, and no sophistry is left unexposed."—*The Rock.*

Science and Scripture.

SCIENCE ELUCIDATIVE OF SCRIPTURE, AND NOT ANTAGONISTIC TO IT; being a Series of Essays on—1. Alleged Discrepancies; 2. The Theory of the Geologists and Figure of the Earth; 3. The Mosaic Cosmogony; 4. Miracles in general—Views of Hume and Powell; 5. The Miracle of Joshua—Views of Dr. Colenso: The Supernaturally Impossible; 6. The Age of the Fixed Stars, &c. By Prof. J. R. YOUNG. Fcap. 5*s*. cl.

Geology.

A CLASS-BOOK OF GEOLOGY. Consisting of "Physical Geology," which sets forth the Leading Principles of the Science; and "Historical Geology," which treats of the Mineral and Organic Conditions of the Earth at each successive epoch, especial reference being made to the British Series of Rocks. By RALPH TATE. With more than 250 Illustrations. Fcap. 8vo, 5*s*. cloth.

Practical Philosophy.

A SYNOPSIS OF PRACTICAL PHILOSOPHY. By Rev. JOHN CARR, M.A., late Fellow of Trin. Coll., Camb. 18mo, 5*s*. cl.

Mollusca.

A MANUAL OF THE MOLLUSCA; being a Treatise on Recent and Fossil Shells. By Dr. S. P. WOODWARD, A.L.S. With Appendix by RALPH TATE, A.L.S., F.G.S. With numerous Plates and 300 Woodcuts. 3rd Edition. Cr. 8vo, 7*s*. 6*d*. cloth.

Clocks, Watches, and Bells.

RUDIMENTARY TREATISE on CLOCKS, and WATCHES, and BELLS. By Sir EDMUND BECKETT, Bart. (late E. B. Denison), LL.D., Q.C., F.R.A.S. Sixth edition, revised and enlarged. Limp cloth (No. 67, Weale's Series), 4*s*. 6*d*.; cloth bds. 5*s*. 6*d*.

"As a popular and practical treatise it is unapproached."—*English Mechanic.*
"The best work on the subject probably extant. The treatise on bells is undoubtedly the best in the language."—*Engineering.*
"The only modern treatise on clock-making."—*Horological Journal.*

Grammar of Colouring.

A GRAMMAR OF COLOURING, applied to Decorative Painting and the Arts. By GEORGE FIELD. New edition, enlarged. By ELLIS A. DAVIDSON. With new Coloured Diagrams and Engravings. 12mo, 3*s*. 6*d*. cloth.

"The book is a most useful *résumé* of the properties of pigments."—*Builder.*

Pictures and Painters.

THE PICTURE AMATEUR'S HANDBOOK AND DICTIONARY OF PAINTERS: A Guide for Visitors to Picture Galleries, and for Art-Students, including methods of Painting, Cleaning, Re-Lining, and Restoring, Principal Schools of Painting, Copyists and Imitators. By PHILIPPE DARYL, B.A. Cr. 8vo, 3*s*. 6*d*. cl.

Woods and Marbles (Imitation of).

SCHOOL OF PAINTING FOR THE IMITATION OF WOODS AND MARBLES, as Taught and Practised by A. R. and P. VAN DER BURG, Directors of the Rotterdam Painting Institution. Illustrated with 24 full-size Coloured Plates; also 12 Plain Plates, comprising 154 Figures. Folio, 2*l*. 12*s*. 6*d*. bound.

Delamotte's Works on Illumination & Alphabets.

A PRIMER OF THE ART OF ILLUMINATION; for the use of Beginners: with a Rudimentary Treatise on the Art, Practical Directions for its Exercise, and numerous Examples taken from Illuminated MSS., printed in Gold and Colours. By F. DELAMOTTE. Small 4to, 9s. Elegantly bound, cloth antique.

"The examples of ancient MSS. recommended to the student, which, with much good sense, the author chooses from collections accessible to all, are selected with judgment and knowledge, as well as taste."—*Athenæum.*

ORNAMENTAL ALPHABETS, ANCIENT and MEDIÆVAL; from the Eighth Century, with Numerals; including Gothic, Church-Text, German, Italian, Arabesque, Initials, Monograms, Crosses, &c. Collected and engraved by F. DELAMOTTE, and printed in Colours. New and Cheaper Edition. Royal 8vo, oblong, 2s. 6d. ornamental boards.

"For those who insert enamelled sentences round gilded chalices, who blazon shop legends over shop-doors, who letter church walls with pithy sentences from the Decalogue, this book will be useful."—*Athenæum.*

EXAMPLES OF MODERN ALPHABETS, PLAIN and ORNA-MENTAL; including German, Old English, Saxon, Italic, Perspective, Greek, Hebrew, Court Hand, Engrossing, Tuscan, Riband, Gothic, Rustic, and Arabesque, &c., &c. Collected and engraved by F. DELAMOTTE, and printed in Colours. New and Cheaper Edition. Royal 8vo, oblong, 2s. 6d. ornamental boards.

"There is comprised in it every possible shape into which the letters of the alphabet and numerals can be formed."—*Standard.*

MEDIÆVAL ALPHABETS AND INITIALS FOR ILLUMI-NATORS. By F. DELAMOTTE. Containing 21 Plates, and Illuminated Title, printed in Gold and Colours. With an Introduction by J. WILLIS BROOKS. Small 4to, 6s. cloth gilt.

THE EMBROIDERER'S BOOK OF DESIGN; containing Initials, Emblems, Cyphers, Monograms, Ornamental Borders, Ecclesiastical Devices, Mediæval and Modern Alphabets, and National Emblems. Collected and engraved by F. DELAMOTTE, and printed in Colours. Oblong royal 8vo, 1s. 6d. in ornamental boards.

Wood-Carving.

INSTRUCTIONS in WOOD-CARVING, for Amateurs; with Hints on Design. By A LADY. In emblematic wrapper, handsomely printed, with Ten large Plates, 2s. 6d.

"The handicraft of the wood-carver, so well as a book can impart it, may be learnt from 'A Lady's' publication."—*Athenæum.*

Popular Work on Painting.

PAINTING POPULARLY EXPLAINED; with Historical Sketches of the Progress of the Art. By THOMAS JOHN GULLICK, Painter, and JOHN TIMBS, F.S.A. Fourth Edition, revised and enlarged. With Frontispiece and Vignette. In small 8vo, 6s. cloth.

** This Work has been adopted as a Prize-book in the Schools of Art at South Kensington.

"Contains a large amount of original matter, agreeably conveyed."—*Builder.*

"Much may be learned, even by those who fancy they do not require to be taught, from the careful perusal of this unpretending but comprehensive treatise."—*Art Journal.*

AGRICULTURE, GARDENING, ETC.

Youatt and Burn's Complete Grazier.

THE COMPLETE GRAZIER, and FARMER'S and CATTLE-BREEDER'S ASSISTANT. A Compendium of Husbandry. By WILLIAM YOUATT, ESQ., V.S. 12th Edition, very considerably enlarged, and brought up to the present requirements of agricultural practice. By ROBERT SCOTT BURN. One large 8vo. volume, 860 pp. with 244 Illustrations. 1*l.* 1*s.* half-bound.

" The standard and text-book, with the farmer and grazier."—*Farmer's Magazine.*
"A treatise which will remain a standard work on the subject as long as British agriculture endures."—*Mark Lane Express.*

History, Structure, and Diseases of Sheep.

SHEEP; THE HISTORY, STRUCTURE, ECONOMY, AND DISEASES OF. By W. C. SPOONER, M.R.V.C., &c. Fourth Edition, with fine engravings, including specimens of New and Improved Breeds. 366 pp., 4*s.* cloth.

Production of Meat.

MEAT PRODUCTION. A Manual for Producers, Distributors, and Consumers of Butchers' Meat. Being a treatise on means of increasing its Home Production. Also comprehensively treating of the Breeding, Rearing, Fattening, and Slaughtering of Meat-yielding Live Stock; Indications of the Quality; Means for Preserving, Curing, and Cooking of the Meat, etc. By JOHN EWART. Numerous Illustrations. Cr. 8vo, 5*s.* cloth.

" A compact and handy volume on the meat question, which deserves serious and thoughtful consideration at the present time."—*Meat and Provision Trades' Review.*

Donaldson and Burn's Suburban Farming.

SUBURBAN FARMING. A Treatise on the Laying Out and Cultivation of Farms adapted to the produce of Milk, Butter and Cheese, Eggs, Poultry, and Pigs. By the late Professor JOHN DONALDSON. With considerable Additions, Illustrating the more Modern Practice, by R. SCOTT BURN. With Illustrations. Crown 8vo, 6*s.* cloth.

Modern Farming.

OUTLINES OF MODERN FARMING. By R. SCOTT BURN. Soils, Manures, and Crops—Farming and Farming Economy—Cattle, Sheep, and Horses—Management of the Dairy, Pigs, and Poultry—Utilisation of Town Sewage, Irrigation, &c. New Edition. In 1 vol. 1250 pp., half-bound, profusely illustrated, 12*s.*

"There is sufficient stated within the limits of this treatise to prevent a farmer from going far wrong in any of his operations."—*Observer.*

Amateur Farming.

THE LESSONS of MY FARM : a Book for Amateur Agriculturists, being an Introduction to Farm Practice, in the Culture of Crops, the Feeding of Cattle, Management of the Dairy, Poultry, Pigs, &c. By R. SCOTT BURN. With numerous Illus. Fcp. 6*s.* cl.

"A complete introduction to the whole round of farming practice."—*John Bull.*

The Management of Estates.

LANDED ESTATES MANAGEMENT : Treating of the Varieties of Lands, Peculiarities of its Farms, Methods of Farming, the Setting-out of Farms and their Fields, Construction of Roads, Fences, Gates, and Farm Buildings, of Waste or Unproductive Lands, Irrigation, Drainage, Plantation, &c. By R. Scott Burn. Numerous Illustrations. Second Edition. 12mo, 3s. cloth.

"A complete and comprehensive outline of the duties appertaining to the management of landed estates."—*Journal of Forestry.*
" A very useful vade-mecum to such as have the care of land."—*Globe.*

The Management of Farms.

OUTLINES OF FARM MANAGEMENT, and the Organization of Farm Labour. Treating of the General Work of the Farm, Field, and Live Stock, Details of Contract Work, Specialties of Labour, Economical Management of the Farmhouse and Cottage, and their Domestic Animals. By Robert Scott Burn, Author of "Outlines of Modern Farming," &c. With numerous Illustrations, 12mo, 3s. cloth boards. [*Just published.*

Management of Estates and Farms.

LANDED ESTATES AND FARM MANAGEMENT. By R. Scott Burn, Author of "Outlines of Modern Farming," Editor of "The Complete Grazier," &c. With Illustrations. Consisting of the above Two Works in One vol., 6s. half-bound.
[*Just published.*

Kitchen Gardening.

KITCHEN GARDENING MADE EASY. Showing how to prepare and lay out the ground, the best means of cultivating every known Vegetable and Herb, with cultural directions for the management of them all the year round. By George M. F. Glenny. With Illustrations, 12mo, 2s. cloth boards.

" As a guide to hardy kitchen gardening, this book will be found trustworthy and useful."—*North British Agriculturist.*

Culture of Fruit Trees.

FRUIT TREES, the Scientific and Profitable Culture of. From the French of Du Breuil, revised by Geo. Glenny. 187 Cuts. 12mo, 4s. cloth.

Good Gardening.

A PLAIN GUIDE TO GOOD GARDENING ; or, How to Grow Vegetables, Fruits, and Flowers. With Practical Notes on Soils, Manures, Seeds, Planting, Laying-out of Gardens and Grounds, &c. By S. Wood. Third Edition, with considerable Additions, &c., and numerous Illustrations. Cr. 8vo, 5s. cloth.

" A very good book, and one to be highly recommended as a practical guide. The practical directions are excellent."—*Athenæum.*

Gainful Gardening.

MULTUM-IN-PARVO GARDENING; or, How to make One Acre of Land produce £620 a year, by the Cultivation of Fruits and Vegetables ; also, How to Grow Flowers in Three Glass Houses, so as to realise £176 per annum clear Profit. By Samuel Wood. 3rd Edition, revised. Cr. 8vo, 2s. cloth.

" We are bound to recommend it as not only suited to the case of the amateur and gentleman's gardener, but to the market grower."—*Gardener's Magazine.*

Bulb Culture.

THE BULB GARDEN, or, How to Cultivate Bulbous and Tuberous-rooted Flowering Plants to Perfection. A Manual adapted for both the Professional and Amateur Gardener. By SAMUEL WOOD, Author of "Good Gardening," etc. With Coloured Illustrations and Wood Engravings. Cr. 8vo, 3s. 6d. cloth.

"The book contains practical suggestions as to the arrangement of the flowers, and the growth of flower-roots for the trade, as well as for amusement."—*Saturday Review.*

Tree Planting.

THE TREE PLANTER AND PLANT PROPAGATOR: Being a Practical Manual on the Propagation of Forest Trees, Fruit Trees, Flowering Shrubs, Flowering Plants, Pot Herbs, &c. Numerous Illustrations. By SAMUEL WOOD. 12mo, 2s. 6d. cloth.

Tree Pruning.

THE TREE PRUNER: Being a Practical Manual on the Pruning of Fruit Trees. Including also their Training and Renovation, with the best Method of bringing Old and Worn-out Trees into a state of Bearing; also treating of the Pruning of Shrubs, Climbers, and Flowering Plants. With numerous Illustrations. By SAMUEL WOOD. 12mo, 2s. 6d. cloth.   [*Just published.*

Tree Planting, Pruning, & Plant Propagation.

THE TREE PLANTER, PROPAGATOR, AND PRUNER. By SAMUEL WOOD, Author of "Good Gardening," &c. Consisting of the above Two Works in One Vol., 5s. half-bound.

Potato Culture.

POTATOES, HOW TO GROW AND SHOW THEM: A Practical Guide to the Cultivation and General Treatment of the Potato. By JAMES PINK. With Illustrations. Cr. 8vo, 2s. cl.

"A well written little volume. The author gives good practical instructions under both divisions of his subject."—*Agricultural Gazette.*

Hudson's Tables for Land Valuers.

THE LAND VALUER'S BEST ASSISTANT: being Tables, on a very much improved Plan, for Calculating the Value of Estates. With Tables for reducing Scotch, Irish, and Provincial Customary Acres to Statute Measure, &c. By R. HUDSON, C.E. New Edition, royal 32mo, leather, gilt edges, elastic band, 4s.

Ewart's Land Improver's Pocket-Book.

THE LAND IMPROVER'S POCKET-BOOK OF FORMULÆ, TABLES, and MEMORANDA, required in any Computation relating to the Permanent Improvement of Landed Property. By JOHN EWART, Land Surveyor and Agricultural Engineer. Royal 32mo, oblong, leather, gilt edges, with elastic band, 4s.

Complete Agricultural Surveyor's Pocket-Book.

THE LAND VALUER'S AND LAND IMPROVER'S COMPLETE POCKET-BOOK; consisting of the above two works bound together, leather, gilt edges, with strap, 7s. 6d.

"We consider Hudson's book to be the best ready-reckoner on matters relating to the valuation of land and crops we have ever seen, and its combination with Mr. Ewart's work greatly enhances the value and usefulness of the latter-mentioned . . It is most useful as a manual for reference."—*North of England Farmer.*

"*A Complete Epitome of the Laws of this Country.*"

EVERY MAN'S OWN LAWYER; a Handy-Book of the Principles of Law and Equity. By A BARRISTER. New Edition, much enlarged. With Notes and References to the Authorities. Crown 8vo, cloth, price 6s. 8d. (saved at every consultation).

COMPRISING THE RIGHTS AND WRONGS OF INDIVIDUALS, MERCANTILE AND COMMERCIAL LAW, CRIMINAL LAW, PARISH LAW, COUNTY COURT LAW, GAME AND FISHERY LAWS, POOR MEN'S LAW, THE LAWS OF

BANKRUPTCY—BILLS OF EXCHANGE—CONTRACTS AND AGREEMENTS—COPYRIGHT—DOWER AND DIVORCE—ELECTIONS AND REGISTRATION—INSURANCE—LIBEL AND SLANDER—MORTGAGES—SETTLEMENTS—STOCK EXCHANGE PRACTICE—TRADE MARKS AND PATENTS—TRESPASS, NUISANCES, ETC.—TRANSFER OF LAND, ETC. — WARRANTY—WILLS AND AGREEMENTS, ETC.

Also Law for Landlord and Tenant—Master and Servant—Workmen and Apprentices—Heirs, Devisees, and Legatees — Husband and Wife — Executors and Trustees — Guardian and Ward — Married Women and Infants—Partners and Agents — Lender and Borrower — Debtor and Creditor — Purchaser and Vendor — Companies and Associations —Friendly Societies—Clergymen, Churchwardens—Medical Practitioners, &c. — Bankers — Farmers — Contractors—Stock and Share Brokers—Sportsmen and Gamekeepers—Farriers and Horse-Dealers—Auctioneers, House-Agents—Innkeepers, &c. — Pawnbrokers — Surveyors — Railways and Carriers, &c., &c.

" No Englishman ought to be without this book."—*Engineer.*

"What it professes to be—a complete epitome of the laws of this country, thoroughly intelligible to non-professional readers. The book is a handy one to have in readiness when some knotty point requires ready solution."—*Bell's Life.*

" A concise, cheap, and complete epitome of the English law, so plainly written that he who runs may read, and he who reads may understand."—*Figaro.*

" A useful and concise epitome of the law."—*Law Magazine.*

" Full of information, fitly expressed without the aid of technical expressions, and to the general public will, we doubt not, prove of considerable worth."—*Economist.*

Auctioneer's Assistant.

THE APPRAISER, AUCTIONEER, BROKER, HOUSE AND ESTATE AGENT, AND VALUER'S POCKET ASSISTANT, for the Valuation for Purchase, Sale, or Renewal of Leases, Annuities, and Reversions, and of property generally; with Prices for Inventories, &c. By JOHN WHEELER, Valuer, &c. Fourth Edition, enlarged, by C. NORRIS. Royal 32mo, cloth, 5s.

" A neat and concise book of reference, containing an admirable and clearly-arranged list of prices for inventories, and a very practical guide to determine the value of furniture, &c."—*Standard.*

Auctioneering.

AUCTIONEERS : THEIR DUTIES AND LIABILITIES. By ROBERT SQUIBBS, Auctioneer. Demy 8vo, 10s. 6d. cloth.

House Property.

HANDBOOK OF HOUSE PROPERTY : a Popular and Practical Guide to the Purchase, Mortgage, Tenancy, and Compulsory Sale of Houses and Land ; including the Law of Dilapidations and Fixtures ; with Explanations and Examples of all kinds of Valuations, and useful Information and Advice on Building. By E. L. TARBUCK, Architect and Surveyor. 2nd Edition. 12mo, 3s. 6d. cl.

'We are glad to be able to recommend it."—*Builder.*

" The advice is thoroughly practical."—*Law Journal.*

Bradbury, Agnew, & Co., Printers, Whitefriars, London.

THE TIMBER IMPORTER'S, TIMBER MER-CHANT'S, AND BUILDER'S STANDARD GUIDE: comprising copious and valuable Memoranda for the Use of the Retailer and Builder. By RICHARD E. GRANDY. Second Edition, revised. 3s.; cloth boards, 3s. 6d.

THE ACOUSTICS OF PUBLIC BUILDINGS, a Rudimentary Treatise on; or, the Principles of the Science of Sound applied to the purposes of the Architect and Builder. By T. ROGER SMITH, M.R.I.B.A., Architect. Illustrated. 1s. 6d.

CARPENTRY AND JOINERY, Elementary Principles of, deduced from the Works of the late Professor ROBISON, and THOMAS TREDGOLD, C.E. Illustrated. New Edition, with a Treatise on *Joinery* by E. WYNDHAM TARN, M.A. 3s. 6d.; cloth boards, 4s.

CARPENTRY AND JOINERY. Atlas of 35 Plates to accompany and illustrate the foregoing book. 4to, 6s.; cloth boards, 7s. 6d.

CONSTRUCTION OF ROOFS, Treatise on the, as regards Carpentry and Joinery. Deduced from the Works of ROBISON, PRICE, and TREDGOLD. Illustrated. 1s. 6d.

QUANTITIES AND MEASUREMENTS in Bricklayers', Masons', Plasterers', Plumbers', Painters', Paperhangers', Gilders', Smiths', Carpenters', and Joiners' work. With Rules for, Abstracting and Hints for Preparing a Bill of Quantities, &c. By A. C. BEATON, Surveyor. 1s. 6d.

LOCKWOOD'S BUILDER'S AND CONTRACTOR'S PRICE BOOK for 1881, containing the latest prices of all kinds of Builders' Materials and Labour, and of all Trades connected with Building, &c. Revised and Edited by FRANCIS T. W. MILLER, Architect and Surveyor. 3s. 6d; half-bound, 4s.

ARCHITECTURAL MODELLING IN PAPER, the Art of. By T. A. RICHARDSON, Architect. Plates. 1s. 6d.

VITRUVIUS' ARCHITECTURE. Translated from the Latin by JOSEPH GWILT, F.S.A., F.R.A.S. Plates. 5s.

GRECIAN ARCHITECTURE, an Inquiry into the Principles of Beauty in. By the EARL OF ABERDEEN. 1s.
  ⁎ With "*Vitruvius*," in one vol., half-bound, 6s.

DWELLING-HOUSES, a Rudimentary Treatise on the Erection of, illustrated by a Perspective View, Plans, Elevations, and Sections of a pair of semi-detached Villas, with the Specification, Quantities, and Estimates. By S. H. BROOKS, Architect. New Edition, with Additions. Plates. 2s. 6d.; cloth boards, 3s.

MATHEMATICAL INSTRUMENTS, a Treatise on; in which their Construction, and the Methods of Testing, Adjusting, and Using them are concisely explained. By J. F. HEATHER, M.A. Illustrated. 1s. 6d.

CROSBY LOCKWOOD & CO., 7, STATIONERS' HALL COURT, E.C.

MECHANICAL ENGINEERING, &c.

MATERIALS AND CONSTRUCTION : a Theoretical and Practical Treatise on the Strains, Designing, and Erection of Works of Construction. By FRANCIS CAMPIN, C.E. 3s. ; cloth boards, 3s. 6d. *[Just published.*

THE BOILERMAKERS' ASSISTANT, in Drawing Templating, and Calculating Boiler and Tank Work. By JOHN COURTNEY, Practical Boilermaker. Revised and Edited by D. KINNEAR CLARK, M.I.C.E. 2s. *[Just published.*

SEWING MACHINERY : a Practical Manual of the Sewing Machine, comprising its History and Details of its construction, with full technical directions for the adjusting of Sewing Machines. By J. W. URQUHART, C.E. 2s. ; cloth bds., 2s. 6d.

THE CONSTRUCTION OF GAS-WORKS, and the Manufacture and Distribution of Coal Gas. Originally written by SAMUEL HUGHES, C.E. Sixth Edition, re-written and much Enlarged by WILLIAM RICHARDS, C.E. 4s. 6d. ; cloth bds., 5s.

FUEL, its Combustion and Economy ; being an Abridgment of "Treatise on the Combustion of Coal and the Prevention of Smoke," by C. W. WILLIAMS, A.I.C.E. With extensive additions on Recent Practice in the Combustion and Economy of Fuel—Coal, Coke, Wood, Peat, Petroleum, &c., by D. KINNEAR CLARK, M.I.C.E. 3s. 6d. ; cloth boards, 4s.

MECHANICS, Rudimentary Treatise on. By C. TOMLINSON. Illustrated. 1s. 6d.

STEAM AND THE STEAM ENGINE, Stationary and Portable, an Elementary Treatise on ; being an Extension of Mr. JOHN SEWELL's "Treatise on Steam." By D. KINNEAR CLARK, C.E., M.I.C.E. 3s. 6d. ; cloth boards, 4s.

THE STEAM ENGINE. By Dr. LARDNER. 1s. 6d.

THE STEAM ENGINE, a Treatise on the Mathematical Theory of. By T. BAKER, C.E. Illustrated. 1s. 6d.

STEAM BOILERS : their Construction and Management. By R. ARMSTRONG, C.E. Illustrated. 1s. 6d.

MARINE ENGINES AND STEAM VESSELS, with Practical Remarks on the Screw and Propelling Power. By ROBERT MURRAY, C.E. 7th Edition. 3s. ; cloth boards, 3s. 6d.

MECHANISM, the Elements of ; elucidating the Scientific Principles of the Practical Construction of Machines. With Specimens of Modern Machines, by T. BAKER, C.E. ; and Remarks on Tools, &c., by J. NASMYTH, C.E. Plates. 2s. 6d. ; cloth boards, 3s.

THE BRASSFOUNDER'S MANUAL. By W. GRAHAM. 2s. ; cloth boards, 2s. 6d.

MODERN WORKSHOP PRACTICE as applied to Marine, Land, and Locomotive Engines, Floating Docks, Bridges, Cranes, Ship-building, &c. By J. G. WINTON. 3s. ; cl. bds., 3s. 6d.

THE WORKMAN'S MANUAL OF ENGINEERING DRAWING. By JOHN MAXTON. 3s. 6d. ; cloth boards, 4s.

CROSBY LOCKWOOD & CO., 7, STATIONERS' HALL COURT, E.C.

Lightning Source UK Ltd.
Milton Keynes UK
UKHW02f1913170518
322795UK00014B/677/P